Valentino Braitenberg · Almut Schüz
Cortex: Statistics and Geometry of Neuronal Connectivity

Springer-Verlag Berlin Heidelberg GmbH

Valentino Braitenberg Almut Schüz

Cortex: Statistics and Geometry of Neuronal Connectivity

Second thoroughly revised edition

With 90 Figures

Springer

Prof. Dr. Dr. VALENTINO BRAITENBERG
Max-Planck-Institut für Biologische Kybernetik
Spemannstr. 38
D-72076 Tübingen
and
Institut für Medizinische Psychologie und Verhaltensneurobiologie
Eberhard-Karls-Universität Tübingen
Gartenstr. 29
D-72074 Tübingen

Prof. Dr. ALMUT SCHÜZ
Max-Planck-Institut für Biologische Kybernetik
Spemannstr. 38
D-72076 Tübingen

The first edition was published under the title Anatomy of the Cortex - Statistics and Geometry as Vol. 18 of the series Studies of Brain Function.

ISBN 978-3-662-03735-5 ISBN 978-3-662-03733-1 (eBook)
DOI 10.1007/978-3-662-03733-1

Library of Congress Cataloging-in-Publication Data
Braitenberg, Valentino. Cortex: statistics and geometry of neuronal connectivity / Valentino Braitenberg, Almut Schütz. - 2nd. thoroughly rev. ed. p. cm. "The first edition was published (1991) under the title: Anatomy of the cortex - statistics and geometry as Vol. 18 of the series Studies of brain function" - T. p. verso.
Includes bibliographical references and index.

1. Cerebral cortex - Anatomy. 2. Cytoarchitectonics. 3. Neural circuitry. 4. Neurons. 5. Synapses. I. Schütz, A. (Almut), 1949- . II. Braitenberg, Valentino. Anatomy of the cortex. III: Title. QM455.B64 1998 611'.81-dc21

Originally published by Springer-Verlag Berlin Heidelberg New York in 1998

Cover design: Design & Production, Heidelberg
Typesetting: K & V Fotosatz, Beerfelden

SPIN 10534572 39/3137-5 4 3 2 1 0 - Printed on acid-free paper

To our relatives, the Rodents,
with apologies

Thanks

(from the preface to the first edition)

This book describes a collective effort in which our whole group at the Max-Planck-Institute for Biological Cybernetics in Tübingen was involved. We are unable to formulate separate eulogies for those of our friends who did the laboratory work and for the others who helped us in the analysis, for there was no clear distinction between them. Some of our technical staff co-authored the papers with us, and the others will in due course. All of them, Monika Dortenmann, Annette Münster, Volker Staiger and Claudia Martin-Schubert participated in our discussions and made essential contributions ranging from the invention of staining methods, through measurements at the microscope, to truly original observations, not to mention their attentive search of the literature in which our group was less defective than our (selective rather than exhaustive) quotations would suggest. Conversely, the mathematician (now professor) Günther Palm did not find it beneath him to check on the minutiae of our counts and measurements, or even sometimes to lend his own eye to them. With Ad Aertsen and his cohort involved in the statistical analysis of electrophysiological data the exchange of ideas was continuous, and certainly some ideas arose as common modes of oscillation in a coupled system involving all of us. Some other ideas were not common modes, but unilateral flow: we received much instruction and advice from our mathematically competent colleagues, and over the years Günther Palm's tutoring was especially helpful.

We owe special thanks to Horst Greilich, Bernhard Hellwig and Michaela Schweizer for letting us use some material from their unpublished academic theses.

Several people read parts or all of our text before it was too late to incorporate their criticism. Shirley Würth and Margarete Ghasroldashti were the first linguistic filters and watched over the form of the manuscript. Shirley was strict and British but we have sometimes evaded her dictate. Elisabeth Hanna-Braitenberg was also consulted and gave stylistic advice (hers more American), especially for the first and the last three chapters. These were also read by Ad Aertsen and Hubert Preißl, whose comments referred to our style of thinking, and were gratefully received by us.

But especially we are grateful to our colleagues abroad whom we had asked for advice and criticism. Professor Ed White in Beer Sheva read one chapter and gave us his candid opinion in several sanguine missives full of valuable information. This was in continuation of a discussion we had had before at a symposium where the truth was spotlighted by our opposite beams. Professor Alan Peters in Boston took upon himself the chore of editing the book, which meant adapting some of our idiosyncratic expressions to the established usage in cortical neuroanatomy. We are exceedingly grateful to him and have followed most of his suggestions. Not all of them: approaches from different angles make the reality break open in different ways, and we thought it useful to preserve the rough edges.

Margarete Ghasroldashti and Claudia Martin-Schubert produced, with admirable skill and patience, the camera-ready copy for the publisher. We are particularly grateful to Hubert Preißl for his expert tutoring in these matters.

All of the art work, and much of the photographic work was done by Claudia Martin-Schubert: she grasps ideas and finds the most effective pictorial form.

Thanks

(added to the second edition)

We are grateful to colleagues, students and librarians who, by their continuing requests, prompted us to prepare a second edition. Many thanks to Dr Czeschlik, Dr Lindenborn and Dr Farrar, our friendly mentors at Springer Verlag, who agreed to publish the book again in a more distinguished format.

This time assistance in matters of organization and word processing was provided by Dorothee Kühnel, to whom we owe much gratitude for her friendly patience and great skill.

* * *

We can wholly dismiss as unnecessary the question whether the soul and the body are one: it is as meaningless as to ask whether the wax and the shape given to it by the stamp are one, or generally the matter of a thing and that of which it is the matter.

Aristotle, *On the soul*

Were ideas entirely loose and unconnected, chance alone would join them; and it is impossible the same ideas should fall regularly into complex ones (as they commonly do), without some bond of union among them, some associating quality, by which one idea naturally introduces another.

Hume, *A Treatise on Human Nature*

For the same things can be thought as can be.

Parmenides

I would especially direct attention to that view of the constitution of the intellect which represents it as subject to laws determinate in their character, but not operating by the power of necessity; which exhibits it as redeemed from the dominion of fate, without being abandoned to the lawlessness of chance.

Boole, *An Investigation of the Laws of Thought*

* * *

Contents

Chapter 1

Introduction

In this essay we propose a view of the cerebral cortex from an unusual angle. Our aim was an explanation of cortical structure in terms of functional principles, and, vice versa, a definition of the role of the cortex in the brain derived from its peculiar structure. Our results may be at variance with the ideas of those physiologists who have specialized subsystems of the cortex in mind, but are not in contrast with them, as we hope to show. Our view is akin to one presented by Moshe Abeles in his monographs *Local Cortical Circuits* (1982) and *Corticonics* (1991). It is related to the theory developed by Günther Palm, for many years our partner, in *Neural Assemblies* (1982). It owes much to the ideas which inspire the work of George Gerstein (beginning with the classic: Gerstein, 1962) and especially to the insights gained by our colleague Ad Aertsen and his group in Tübingen (now Freiburg). Our ideas on the cortex find some resonance in a particular brand of Artificial Intelligence (Kohonen, 1977; Hopfield, 1982; Rumelhart et al., 1986), and indeed a diagram representing the dominant idea of that trend, the associative matrix with feedback, was originally proposed as a summary statement of cortical connectivity (Braitenberg, 1974a).

Since our earlier reports (Braitenberg 1974a,b, 1977, 1978a,b, 1986; Schüz, 1976, 1978, 1981a,b, 1986) enough new experimental evidence had accumulated by 1990 in our own laboratory (summarized by Schüz, 1989) and elsewhere to justify a more extended and more confident presentation in the first version of this monograph (Braitenberg and Schüz, 1991). Today, a new edition seems appropriate, not only because the first edition was sold out, but also because a more friendly attitude towards the statistical approach to cortical neuroanatomy seems to prevail now, and several laboratories have produced fresh evidence which complements our own. There are some additions due to work done in our own laboratory and elsewhere in recent years, and some minor corrections, but luckily none of our main tenets had to be taken back.

Briefly, our points are: (1) Most of the synapses in the cortex are involved in the business of associative memory. (2) To make this more efficient, the information in the cortex is thoroughly shuffled: much of the fibre anatomy points to the cortex as a great mixing device. (3) However, information belonging to different contexts is kept locally separate: this is the point of areal diversity. The special structure of each area (different in a quantitative way, rather than in principle) incorporates inborn knowledge about the rules and regularities that are to be expected in the various contexts.

When we first started contemplating the structure of the cerebral cortex, it was a light-hearted exercise in speculative neuroanatomy, of the sort we had applied earlier to other pieces of nerve tissue endowed with a strikingly orderly build. When the rich tangle of neuronal elements in the cerebral cortex is contrasted to the sober wiring of the visual ganglia of the fly, or to the geometric simplicity of the cerebellar cortex (Fig. 1), the impelling question is: can we say anything about it at all? Is it possible to imagine a useful scheme of information handling, whose realization in a neuronal network requires precisely the kind of elements and connections which are found in the telencephalic cortex of mammals?

The scientific community was not, at the time, in favour of such fanciful neuroanatomy. The prevailing notion was that of electronic wiring; the dominant sentiment one of awe in the face of a complexity which would seem forever beyond our conceptual means if interpreted according to that notion.

Moreover, we had all been impressed by a series of insights into the functional neuroanatomy of the visual cortex which had emerged from microelectrode work by Hubel and Wiesel. This seemed to many to point the way to a complete understanding of the cortical level of neuronal computation, whereupon all other ways were held to be at best superfluous.

We were not discouraged by either argument. Our basic hunch was that "wiring" in the conventional sense could not be the answer. Among other things, the enormous amount of structural order which would be embodied in the cortex if every synapse was to be interpreted as a line in an electronic wiring diagram seemed biologically impossible: the genes are an insufficient channel for so much order, and other mechanisms, e.g. early learning, do not seem to have the necessary spatial finesse.

We also felt that the results of Hubel and Wiesel and their colleagues ultimately will bear fruit only if they can be related to some general principle of information handling in the cortex. Without such underpinning they remain essentially phenomenological.

Thus the direction of our thoughts was determined. The problem was to relate the structural features which are typical of the cerebral cortex throughout the mammalian species and throughout the local areal variation, to some operation which is performed there and only there. Because of the widely varying relations of the cortex to input and output systems,

Fig. 1. Various degrees of order and disorder in nerve tissue. *Upper* tangential section through layer IVa of monkey area 17; *middle* tangential section through the (curved) layer of L_4 collaterals in the lamina ganglionaris of the fly. *Lower* tangential section through the lowest level of the molecular layer of the cerebellar cortex of the mouse. Axons of basket cells run vertically, parallel fibres horizontally

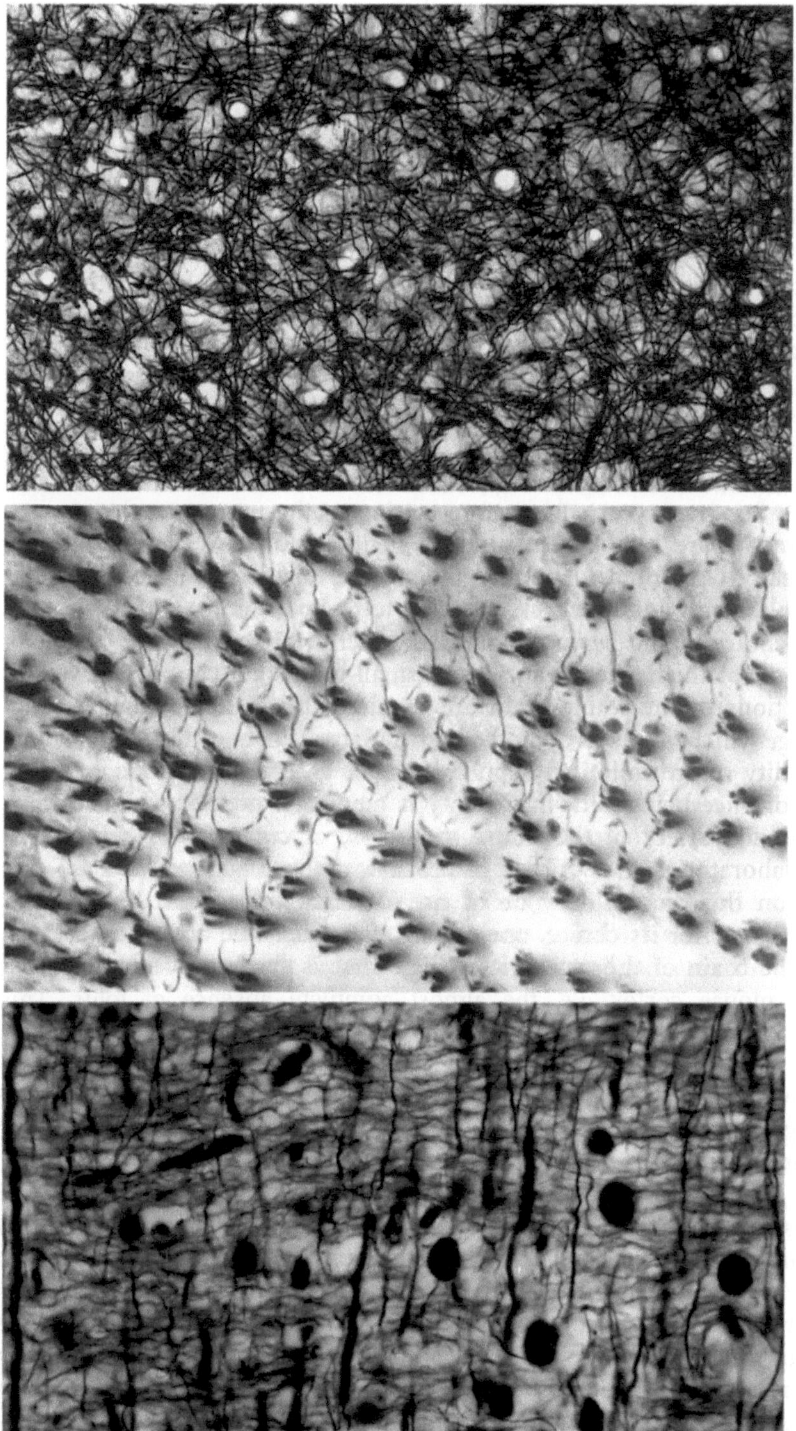

Fig. 1

it had to be an operation of a very basic kind, sufficiently general to be applicable to visual input, auditory input, tactile input, to the generation of motor output and even to the items which circulate through the speech areas of the human cortex.

It was also clear from the outset that the ideas we were about to generate about information handling in the cortex would have a statistical touch. Once individual synapses and fibres are recognized as providing a picture which is far too detailed to be biologically realistic, the correct description is one in terms of *populations* of synapses and fibres. If the genetic planning of the cortex and its subsequent moulding through experience are in terms of statistical rules, the appropriate neuroanatomy must also be in that language. Our own essay follows a tradition inaugurated by Sholl, 1956, Bok, 1959 and Cragg, 1967.

We will introduce our reader to the thoughts that came to us in the first playful phase, and to the hypotheses which determined our laboratory work for many years. But before getting to the more speculative part, we cannot spare him the quantitative neuroanatomy we have produced, because it is essential for our argument. He may skip the technical details, unless he wants to check on us, or to compare notes.

The choice of the *white mouse* as the victim of most of our histology is easily explained. The advantages of small brains are clear whenever serial sections of the whole brain are required. In studies involving Golgi staining, which is, as everybody knows, largely a question of luck, the availability of an unlimited supply of brains is comforting, the breeding habits of mice being such that plenty is more of a problem than scarcity. The albino variety of the house mouse is, of course, very familiar in many laboratories, not only of neuroanatomy, and so the facts we collected on this species may be of use to many colleagues. Two further good reasons for its choice, one certain, the other one not quite so clear, lie in the brain of the mouse itself. The first is the lyssencephalic layout of the mouse cortex, i.e. the almost complete absence of folds. This makes it very easy to obtain sections in one of the intrinsic coordinates of the cortex (horizontal or vertical) by orienting the section appropriately with respect to the surface of the hemisphere: with the elaborate folding patterns in larger "gyrencephalic" mammals the coordinates of the cortex frequently evade the macroscopical order.

The other reason, of long standing in neuroanatomical folklore, is the supposedly more homogeneous build of the mouse cortex compared to the variety of cortical areas as described, e.g. in the human brain. This idea was often extended (already by Cajal) to include the idea of a lesser variety of neuronal morphology in rodents and even the absence of some distinct neuronal types which are found in more distinguished species (Valverde, 1991). In our experience, this belittling of the mouse brain is not justified. Golgi preparations of the mouse cortex provide an impressive display both of architectonic variety (Rose, 1929) and of neuronal types (Lorente de Nó, 1922, 1938), fully sufficient to illustrate all the problems that face us when we consider brains of primates or of carnivores. Moreover, if there is good reason to consider the behaviour and the brain of the mouse as being at or near one extreme of the mammalian variety, and the human behaviour and brain at the other, what an opportunity to discover the common features of two very different editions of the same apparatus, the structural invariants that can be supposed to contain the key to the basic operations of the cortex! And the common features of the cortex in all mammals are indeed striking. At a recent meeting on the cortex, well attended by neuroanatomists, when the role of the cortex in the noblest cognitive human performances was mentioned in the discussion, we projected slides of the bovine cortex and nobody noticed. The trick would have worked equally well with appropriately chosen Golgi pictures of the mouse cortex. The mouse may teach us not to aim too high when looking for the operations typical for the telencephalic level.

Finally, a very good reason for choosing the mouse is the good company in which one finds oneself. There are three sources which make life easy for the newcomer in mouse-cortical neuroanatomy. The atlas by Sidman, Angevine and Pierce (1971) lays the groundwork. Its plates are of such quality that some low magnification cyto- and myeloarchitectonic histology can be done directly on them. The next monumental collection of murine cortex studies is contained in some of the volumes *Cerebral Cortex* edited by Peters and Jones (1984 onward). Much of what is said there by various authors about the cortices of other animals is also very valuable to us, since we do not believe that interspecific differences override the fundamental identity of the cortical network. Most importantly, the book *Cortical Circuits* by Ed White (1989) made available in compact form a large amount of research on the mouse cortex which the author has done over the years. Since White offers a good review of the literature, and since his point of view is different from ours, we are very grateful to him for the opportunity of referring our readers to his book when we would otherwise have owed them a lengthy discussion.

How findings obtained from the mouse cortex relate to the cortical structure in other mammalian species is a topic which we will take up in chapter 34.

Chapter 2

Where is the Cortex?

It is fairly easy to provide a description of the cortical histology which could serve as an adequate criterion for the identification of almost any piece of cortex from any mammal, and for the exclusion of almost any other piece of the brain. Thus the structural traits of the cortex in general seem to be well defined. Yet it is quite a different matter to define the limits of the cortical tissue in any particular brain, in both the topographic sense, and in the sense of a distinction of the extreme varieties of cortical build which are still to be considered as cortex from similar structures which are definitely no longer part of it. This problem is, of course, common to many biological contexts, where the definition of a certain type (of tree, of sense organ, of society, of linguistic syntax) becomes quite elusive when put to the test of a decision procedure capable of recognizing every instance of it.

Tentatively, we define the cortex as that part of the grey substance in the anterior part of the brain which has a planar layout, with a fairly homogeneous appearance in every direction of the plane but with an ordered structure at right angles to it, the so-called layers. The sequence of layers, together with the fact that the fibre masses connecting the cortex to other parts of the brain are mostly situated on one side, define an up-down direction perpendicular to the cortical plane, the white substance being considered "below" the cortex. This is of course quite arbitrary, because even in animals whose cortex is not rich in folds, some of the cortex is tucked into the skull in such a way that the white substance comes to lie above it. Inside-outside would have been a more adequate name for the "vertical" coordinate of the cortex.

We shall see later how the analysis of intracortical connections provides even better criteria for setting apart the telencephalic cortex from other parts of the brain. If the definition of the cortex is to be functionally meaningful, the peculiar character of its internal wiring should, of course, be the main guideline for its anatomical delimitation.

Unfortunately, both the macro- and the microscopic criteria fail us when we examine some of the marginal regions where the cortical grey substance goes over without interruption, i.e. without interposed regions of white substance, into other pieces of grey substance not considered cortex in the conventional terminology. We shall not attempt to make an exact definition of the borders there, nor will we even consider the question whether such a definition is at all possible.

We had at our disposal series of sections through adult mice brains cut in the three conventional directions, frontal, horizontal and sagittal. The staining was Nissl (Cresyl violet on paraffin or frozen sections) and, for myelinated fibres, the procedure proposed by Werner (1986), which in our experience is one of the most reliable myelin stains on the market today, the haematoxylin of the old days no longer being available.

For questions of terminology, the atlas by Sidman, Angevine and Pierce (1971) was our guide. We shall refer to this work as the SAP-atlas.

Forward Delimitation of the Cortex Towards the Olfactory Bulb. By any stretch of the imagination, it is impossible to describe the olfactory bulb as a continuation, with variations, of the layered structure of the cortex (Fig. 2), even if one might be tempted to do so on the basis of the alleged archaic relatedness of these two parts of the forebrain. Both the granular cell layers and the mitral cell layer of the bulb terminate abruptly with no transition into any of the cortical layers. The glomerular layer is even more clearly separated from the rest of the forebrain. Moreover, one is completely at a loss when trying to follow the coordinates of the cortex – the cortical plane and the vertical direction – into the olfactory bulb. If anything, the bulb is a mushroom-shaped excrescence whose axis is perpendicular to the surface of the telencephalic hemisphere near its anterior pole.

What is not so easy to establish is the border between the olfactory bulb and the so-called olfactory nucleus, whose cell masses penetrate the core of the mushroom stem, but without merging with the cell layers of the bulb proper (Fig. 2).

The olfactory nucleus itself, although well marked by its dense cell layer in the Nissl picture, is definitely a continuation of the cortex. The entire complex pyriform cortex, olfactory nucleus, olfactory tubercle (Figs. 2, 3) is well aligned with the cortical plane, or with the upper layers of the cortex, to be more precise. We would not know where to mark the end of the cortex and the beginning of something else, in spite of the misleading term "olfactory nucleus". This is an important consideration for those who like to think of the global operation of the cortex in analogy with some so-called "neural network" models. If the cortex forms a unit characterized by its own peculiar statistics of internal connections, its delimitation from neighbouring grey matter is probably a question of a quantitative rather than qualitative kind.

Continuation of the Cortex into the Hippocampus and Beyond. The cortex of the hippocampus definitely continues the cortical sheet (Fig. 2). If the hemisphere is stylized as the surface of a cone with its tip at the olfactory bulb (and its base missing), the hippocampus occupies most of the posterior margin of the conical shell (Fig. 5). The continuity of the hippocampal layers with those of the adjoining part of the cortex can be clearly visua-

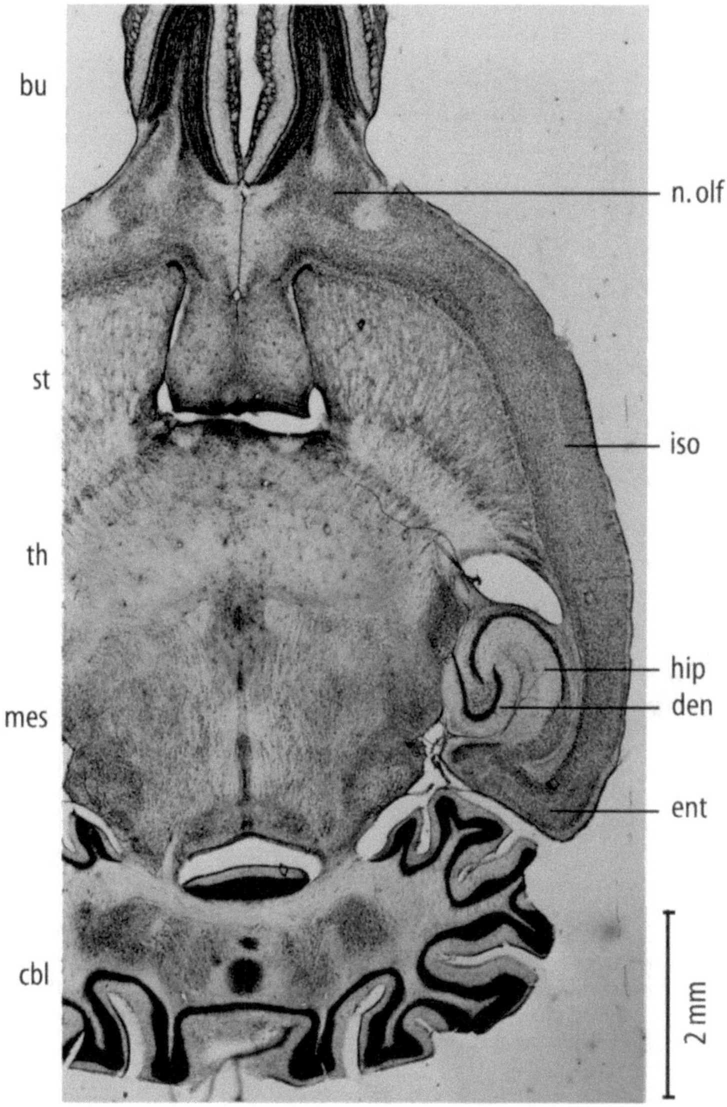

Fig. 2. Horizontal Nissl-stained section through the mouse brain. Demarcation of the olfactory bulb from the cortex. Transition of the cortex into the hippocampus. *bu* olfactory bulb; *st* striatum; *th* thalamus; *mes* mesencephalon; *cbl* cerebellum; *n.olf* olfactory nucleus; *iso* isocortex; *hip* hippocampus; *den* dentate gyrus; *ent* entorhinal cortex

lized, as we shall demonstrate in Chapter 32. Thus there is every reason to include the hippocampal cortex with the rest of the cortex.

It is a question, however, whether the dentate gyrus of the hippocampus should be included. It, too, has a planar architecture, but the plane (or better, bent surface) in which its cellular elements are arranged is not

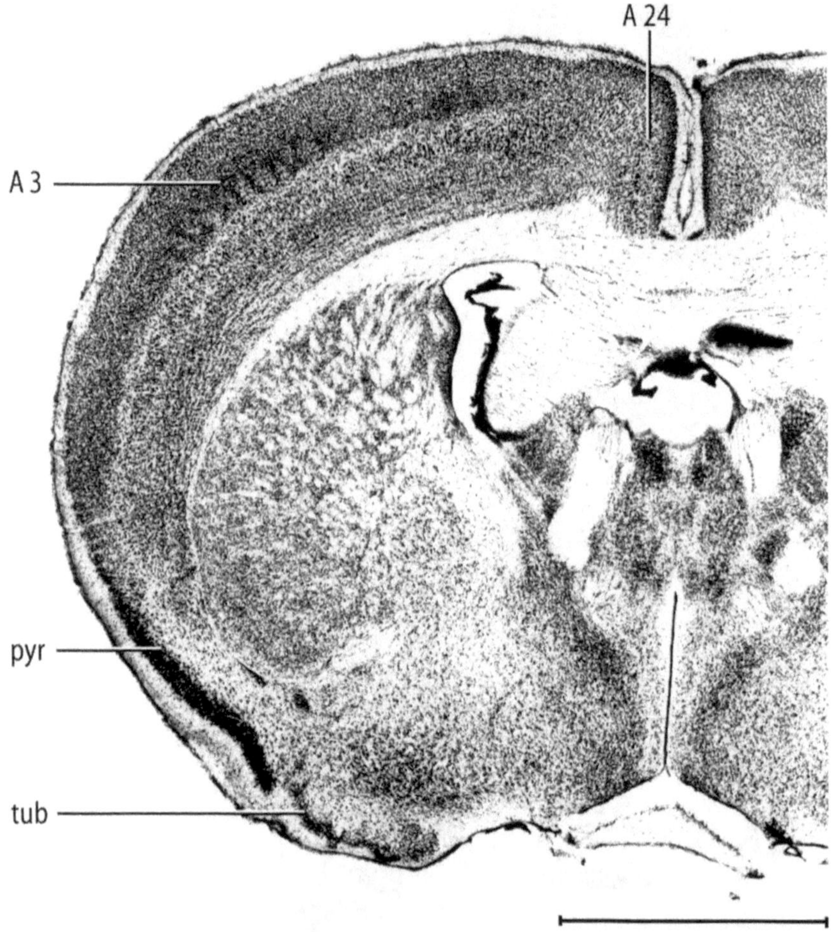

Fig. 3. Frontal section through the mouse brain, left hemisphere. Nissl stain on a thick section, to enhance areal differences. *A24* anterior cingulate cortex (area 24 of the Caviness map); *A3* the "barrel field" (area 3); *pyr* pyriform cortex; *tub* olfactory tubercle. (cf. Figs. 73 and 74). Bar: 2 mm

continuous with that of the rest of the cortex. Its relation to the adjoining part of the hippocampus can best be described as that of two planes at right angles to each other. Still, it makes no sense to assign to the dentate gyrus the dignity of a separate organ, particularly since its fibre connections with the hippocampus are intimate and exclusive.

The telencephalic grey substance, on the other hand, ceases to be cortex in the amygdala (Fig. 4). Here the planar order of the cortex, so striking in the hippocampal cortex, gradually subsides, and there is a similar gradual transition between the amygdala and the pyriform cortex in front of it. There are no fibre layers or connective tissue gaps to demarcate the cortex in this region. The SAP atlas decides that part of the amygdala is

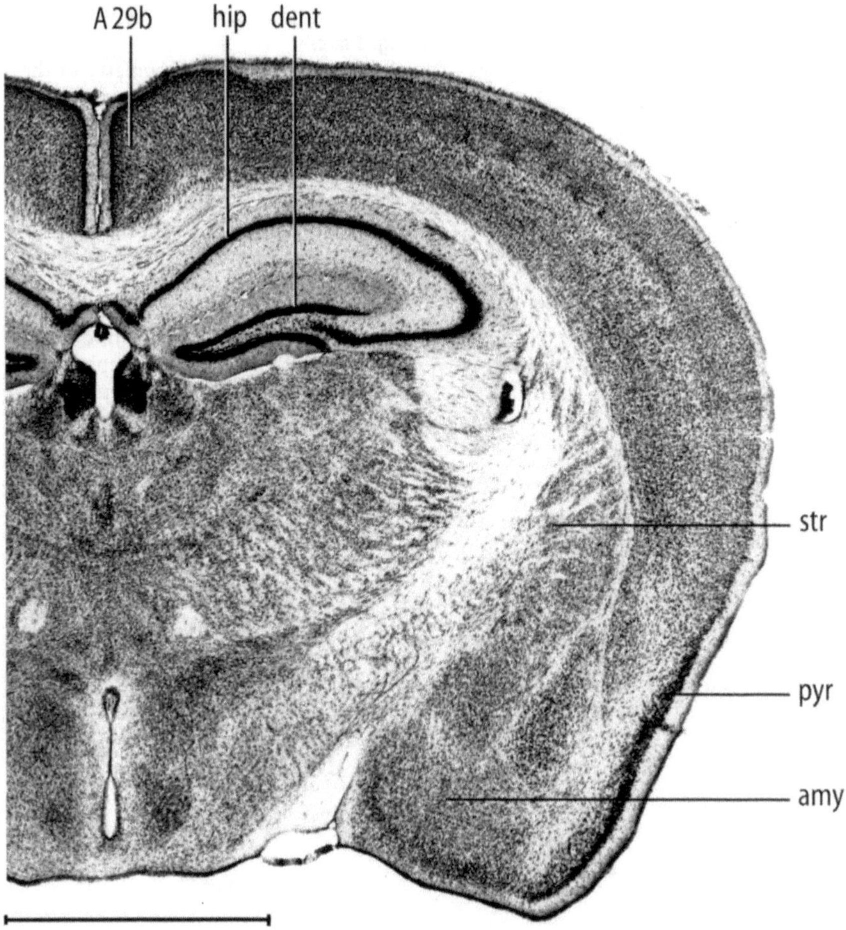

Fig. 4. Frontal section through the mouse brain, right hemisphere. The section is posterior to that of Fig. 3. Nissl stain on a thick section, to enhance areal differences. *A29b* posterior cingulate cortex (cf. Fig. 65); *hip* pyramidal layer of the hippocampus; *dent* granular layer of the dentate fascia; *str* striatum, *pyr* pyriform cortex; *amy* the amygdala region. Bar: 2 mm

cortical (n. amygdaloideus corticalis) and another part is not (n.amygdaloideus basalis, lateralis etc.). The distinction between a cortical and a non-cortical part of the amygdala can also be made considering the shapes of neuronal ramifications in Golgi pictures (McDonald, 1992).

It is noteworthy that in the mouse brain, starting from the cortical grey substance, one could leave the cortex through the amygdala and, without ever crossing white substance, could reach the basal ganglia and from there the septum and the hypothalamus. We do not know enough about the degree of functional independence of the various suborgans of the brain to develop an intuition for the possible significance of the *grey*

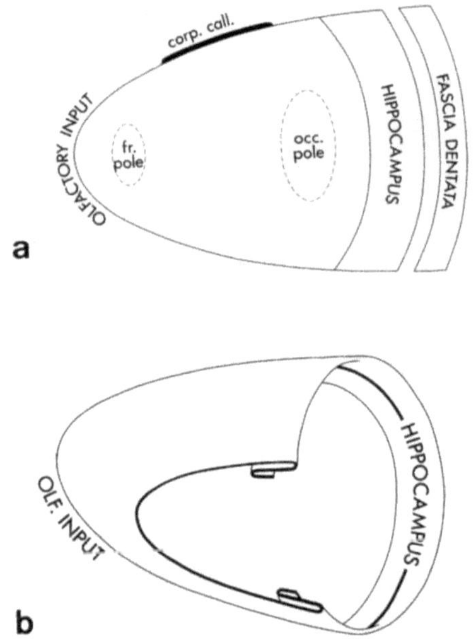

Fig. 5 a–c.
Cartoons showing the geometry of the attachment of the hippocampus to the rest of the cortex. **a** The roughly triangular shape of the unfolded mouse cortex, with the hippocampus opposite the frontal olfactory region. The dentate gyrus (fascia dentata) is a separate piece of cortex. **b** The cortex bent into a roughly conical shape. The posterior margin, the hippocampus, is bent inward and then outward again. **c** A more schematic view of the telencephalic hemisphere with its medial opening, and a more realistic view of the hippocampal formation. The subiculum continues the lower layers of the cortex (presubiculum). (Braitenberg and Schüz, 1983)

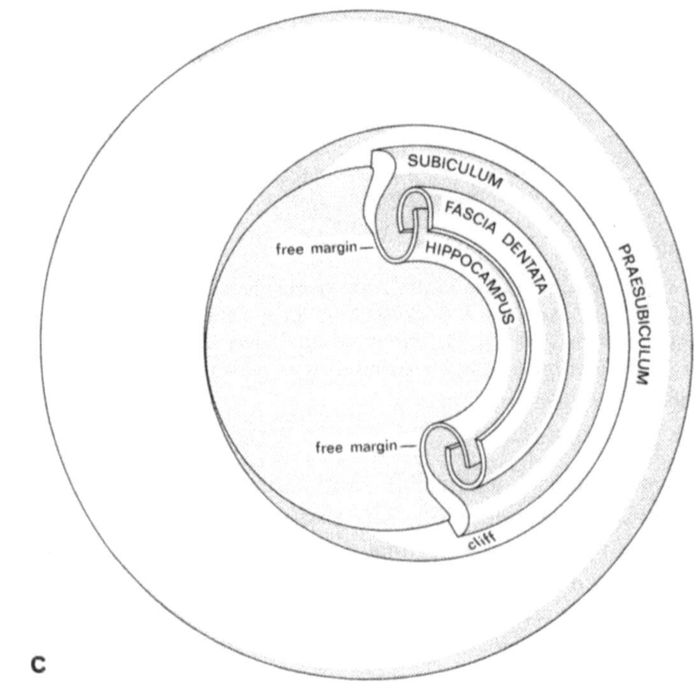

continuity which some regions enjoy with each other, while others communicate only through fibre bundles.

Medially the margin of the cortex is well defined in the dorsal part of the hemisphere by the interhemisphere gap (Figs. 3, 4). Below the cortex the margin is well defined everywhere, even where there is almost no white substance between the cortex and the striatum, the latter having a definitely non-cortical appearance in both its fibre and cell architecture (Fig. 3).

Chapter 3

Analysis of the Neuropil: Preliminary Remarks

The basic tenet in our treatment of cortical histology is the following. We believe that important insights in the nature of the cortical network can be gained by the quantitative comparison of statistical measures such as the density of various tissue elements (number of neurons, synapses, dendritic spines per unit volume, total length of axons and dendrites per unit volume, average range and distribution of axons and dendrites around the neurons) even if the detailed pattern of the connections is not known. Before we can show this (Chap. 13), we must present our measurements in some detail. Because of the importance of densities in our quantitative description, some preliminary technical comments are in order, referring to shrinkage which may arise in the histological procedure, and to the general difficulty of assessing densities in the volume from measurements which are by necessity performed on quasi two-dimensional sections.

3.1 Tissue Shrinkage (of Interest Mainly to Neuroanatomists)

The preparation of the tissue for histology may involve dehydration, precipitation of proteins, polymerization and other processes that are poorly understood. The effect in almost all cases is a change in volume, usually shrinkage, although some of these processes (e.g. fixation with Osmium) may actually add to the volume of the original specimen. Shrinkage or swelling can scarcely be disregarded when the aim is a quantitative assessment of various components of the nerve tissue. These effects are particularly important when the measurements refer to densities, i.e. number of elements per unit volume. In this case a small change in the linear dimensions of the specimen, not at all striking to the eye, when elevated to the third power as it would be in the calculation of the volume, may considerably affect the results.

What is worse, we were often forced to compare quantities that were measured on material prepared with different histological techniques. For example, we measured the density of the axonal feltwork on frozen sections of aldehyde-fixed material and compared this to the measurements of the axonal length of chromated Golgi material and to neuron counts performed on paraffin-embedded, Nissl-stained sections. Here the assessment of shrinkage with the different preparations became imperative.

The problem was felt by many (e.g. Stephan, 1960; Romeis, 1968; Werner and Winkelmann, 1976; Mouritzen Dam, 1979; O'Kusky and Colonnier, 1982; Drenhaus et al, 1986) and was tackled again by Schüz and Palm (1989), who compared the following histological techniques with regard to the tissue shrinkage they produce: (1) the procedure for electron microscopy as proposed by Palay and Chan-Palay (1974); (2) Nissl stain on mounted sections after fixation in 3.7% formaldehyde and embedding in paraffin; (3) Nissl stain on mounted frozen sections (after fixation in 3.7% formaldehyde and impregnation of the tissue with 30% glucose overnight before freezing); (4) Golgi stain according to Colonnier (1964) followed by embedding in celloidin; (5) Golgi stain and frozen sections. (In this case the material had not been impregnated with glucose before freezing).

The procedure was the following. Seven brains were perfused and removed from the skull the next day. Each brain was then trimmed into a rectangular block and then again into three smaller blocks of the same lateral width. Each of these blocks was photographed and then subjected to one of the different methods. The blocks were photographed again at different steps of the procedures, always at the same magnification. By superposition of the pictures it was easy to compare the width of the blocks with that of the ultimate preparations.

The tissue prepared for electron microscopy could be investigated in this way only up to the stage of semithin sections. For the last step the blocks have to be trimmed down to a very small size to produce sections which fit onto the grids used for electron microscopy. They are then treated with lead citrate and uranyl acetate. In order to control possible changes during this last step, the distance between conspicuous structures was measured both on a semithin section in the light microscope and on the directly adjacent thin section in the electron microscope.

The comparison between frozen and celloidin sections of Golgi-stained material was carried out on the two hemispheres of one and the same brain.

It still had to be clarified which amount of shrinkage occurs during perfusion and in-situ fixation, i.e. before the first photographs could be taken. To investigate this, two animals were deeply anaesthezised and their skulls exposed. Two small holes were drilled from above through the skull into the cortex and the skull was then photographed. One of the animals was perfused as for electron microscopy, the other one with 3.7% formaldehyde. The brains were removed the next day and the dorsal surface, showing the lesions made by the drill, was photographed at the same magnification as previously. By superposition of the pictures one could compare the distance between the holes.

The results were the following. When the brain is perfused with aldehydes for electron microscopy or with 3.7% formaldehyde and removed the next day, no changes in size can be detected. In Osmium, there is a clear swelling of the tissue which is, however, compensated by later dehydration and embedding. This means that there is very little change in volume after the whole procedure. This is also true for the last step from the semithin to the thin section.

The most pronounced shrinkage was found in paraffin-embedded Nissl material. The variability was rather high, ranging between a linear factor of 0.67 and 0.81 in different brains. A comparison between the linear dimension of the sections parallel and perpendicular to the edge of the knife showed that the sections were more "shrunken" in the dimension perpendicular to the long axis of the knife, as if the edge had compressed them (without producing folds). Taking this fact into account and averaging over the different brains, one obtains a mean value of 0.75 of linear shrinkage for paraffin-embedded Nissl material.

The shrinkage after Nissl stain on mounted frozen sections was considerably less than on paraffin-embedded material. The linear shrinkage was 0.88.

The Golgi preparations embedded in celloidin were shrunken by a linear factor of 0.89. Surprisingly, there was no difference between celloidin embedded and frozen sections of Golgi-stained material.

The results can be summarized thus. With the methods for electron microscopy the tissue shrinks to 96% of its original volume; with paraffin embedding followed by Nissl staining to 42%; with Nissl staining on frozen sections to 68% and with Golgi staining to 70% of its original volume.

3.2 Inferring the Density in the Volume from the Density in Sections

This problem was long felt by histologists who wanted quantitative data, and many methods were designed to overcome the practical difficulties. In the simplest case, e.g. counting spherical objects all of the same size in the tissue, little more has to be done than to take into account the objects which are only partly contained in the section, and correct the counts accordingly. The problem is much more complicated, and in some cases even unsolvable in principle except by crude approximation, when the objects are of varying size and especially when they have different shapes and orientations in the tissue examined. In such cases the distribution of shapes and sizes would have to be determined first before any reliable correction could be applied to the counts, but this distribution cannot be determined without incurring in the same errors which it is intended to overcome. Think of spherical objects of varying sizes. Some of them are entirely contained in the section and appear as circles with different diameters. But these cannot serve to determine the distribution of sizes, since the sample may be contaminated by other spheres which are only partly contained in the section and whose cross-sections may appear smaller than their real diameter.

These difficulties are very troublesome, and explain in part the widely varying figures which different quantitative neuroanatomists report for the same kind of measurement on the same kind of material. Several authors have given professional treatment to these problems which now go by name of *stereology* (e.g. Weibel, 1969; Konigsmark, 1970; Haug, 1979 a, b; Uylings et al., 1986; Reith and Mayhew, 1988; Clarke, 1993). In some of our own work we availed ourselves of the mathematical advice of Günther Palm (Schüz, 1981 a, b; Schüz and Palm, 1989) who, not content with applying standard methods to our problems, developed in some cases an analysis of his own.

We shall give a brief survey of the methods which we found helpful.

A time-honoured procedure, when counting cells or other entities in the tissue, is to reduce each of these entities to a small, easily identifiable part of them and to count these. The classic example is counting nucleoli instead of whole nerve cells. The diameter of the nucleoli being much smaller than that of the cell body and of the nucleus, we expect them to be only very rarely, if at all, cut in two by the sectioning. One of the sources of double counts being thus practically eliminated, there is another one in those neurons which contain more than one nucleolus, or more than one lump of chromatin that could be interpreted as such.

A clever variation of this method was formalized by Sterio (1984). When the objects in the tissue do not contain any small marker (such as the nucleolus), it is still possible to count only the uppermost, or the lowermost point of these objects contained in the section. This can be done in a very simple way by subtracting from the count obtained in one section the number of those objects that *also* appear in a neighbouring section. Thus one counts only the objects whose upper (or lower, depending on which neighbouring section was used) end is contained in the section. The great advantage of this method is that it works for objects of any size and shape.

Bok (1959) had introduced a similar trick. He counted cells in neighbouring sections of different thickness and then subtracted the count obtained on the thinner section from that on the thicker one. The difference could be taken as the true count on a hypothetical section whose thickness is the thickness of one minus the thickness of the other, all the difficulties with cell bodies or nucleoli split by the knife being the same in the two sections and thus disappearing in the subtraction.

Most of the methods which are used to relate counts of objects in a section of known thickness to the density of these objects in the tissue introduce a correction for those objects that are partly outside of the section. For the formula which is most often applied, Abercrombie (1946) is quoted. Considering the simple case of spherical objects of uniform size (Fig. 6), a section of thickness s contains fragments of these objects provided they are within a region defined by the two planes $d/2$ above the section and $d/2$ below, where d is the diameter of the objects. Thus the

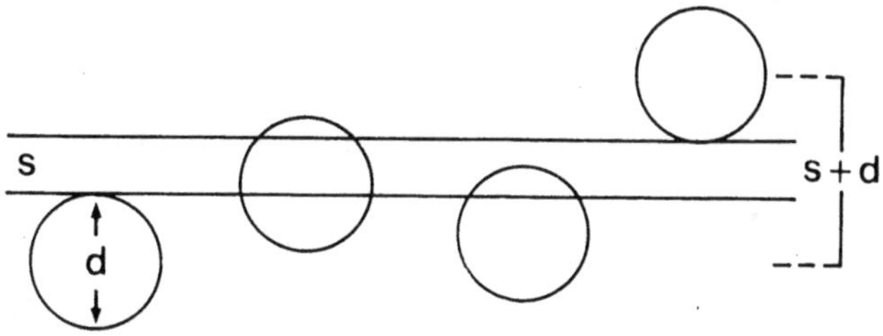

Fig. 6. Particles of size d which are partly contained in a slice of thickness s have centres lying within a virtual slice of thickness $s + d$

volume to which the count is to be related is the surface F of the section multiplied by $s + d$, and the real number N of objects in the volume is obtained from the count N_c in the section by

$$\frac{N}{\text{mm}^3} = \frac{N_c}{F(s+d)} \,. \tag{1}$$

The trouble is that even with spherical objects of uniform size it is not always clear how thin a fragment of the objects can still be recognized in the microscope, and in some cases it is even possible that for mechanical reasons a small cap of a spherical object may fall out of the section.

For objects of different sizes we applied the following modification of the Abercrombie procedure (Schüz, 1981a; Schüz and Palm, 1989). The average apparent diameter of the objects (in our case, neural cell bodies in the cortex) was estimated first. For each set of objects of size d, cut by a section of thickness s (Fig. 7) the apparent diameters δ are distributed according to the distribution depicted on Fig. 7b. The average of the apparent diameters δ can be calculated as the integral of this distribution:

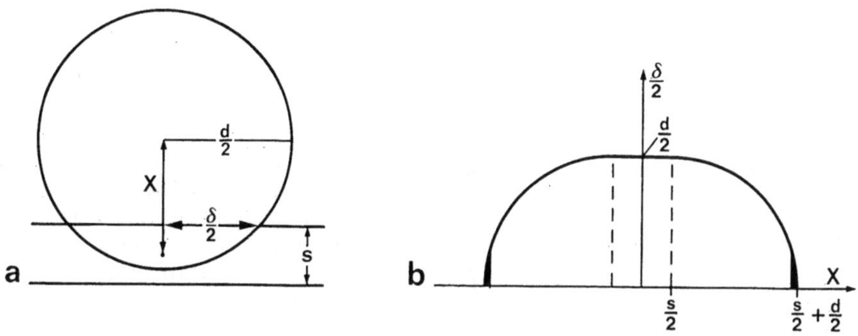

Fig. 7. a A sphere with diameter d is partly contained in a slice of thickness s. x is the distance between the center of the sphere and the middle layer of the slice. δ is the diameter of the part contained in the slice. **b** Dependence of δ on x. This is the distribution of apparent diameters δ if all the positions of particles of uniform size relative to a histological section occur with equal probability. (Schüz, 1981b)

$$\bar{\delta} = \frac{ds + \pi d^2/4}{s + d} \,. \tag{2}$$

Solving this for *d*, we obtain the real diameter from the average of the apparent diameters δ:

$$d = \frac{2}{\pi}(\bar{\delta} - s) + \sqrt{\left[\frac{2}{\pi}(\bar{\delta} - s)\right]^2 + \frac{4}{\pi}\bar{\delta}\, s} \tag{3}$$

It is mathematically not altogether correct, but good enough in practice, to use this formula also in the case of objects of different sizes in order to derive the average real diameter of the objects from their average apparent diameter in the section. To do this exactly, one would have to make unwarranted assumptions about the distribution of their sizes in the tissue.

Counting *synapses* on electron micrographs requires a more complicated analysis. Most synaptic junctions[1] in the cerebral cortex have the shape of roundish plaques of varying sizes (Figs. 13d, 45). Thus, even if we neglect the fact that the plaques may be bent in one or two directions, we must consider their different orientation beside their different position within the section (Fig. 8). One of the consequences is that we cannot directly apply the Abercrombie correction by considering the diameter of the synapse, as we did in the case of the spherical objects.

A flat synaptic junction may have its centre closer to the section than one half its diameter and may still not touch it at all, if its plane is not perpendicular to that of the section. Considering all the possible orientations of the synapses, the virtual thickness of the section (in the sense of

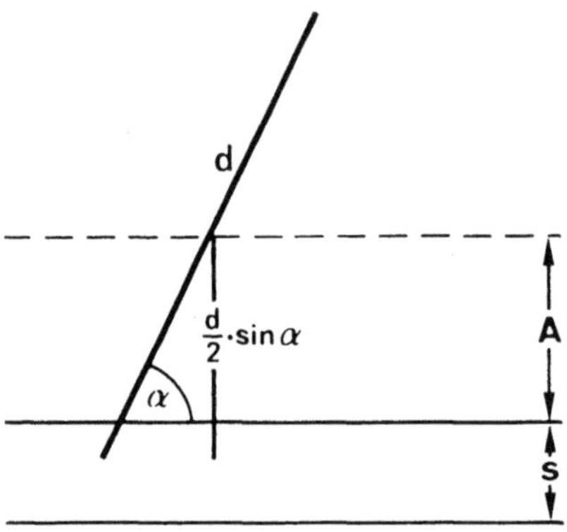

Fig. 8.
The geometrical situation involved in counting disc-shaped objects of diameter *d*, oriented at various angles *a* which are partly contained in a slice of thickness *s*. *A* is the distance of the centre of the disc from the surface of the slice

[1] This is the established term for what an electron microscopist usually takes as evidence for a synapse on electron micrographs, the pre- and postsynaptic membrane thickening. When there is no danger of being misunderstood, we will sometimes say synapse when we should say synaptic junction.

the Abercrombie correction) could be taken as the real thickness s plus twice A (Fig. 8), where A is the average distance of the centre of the synapse from the section.

$$A = \frac{\int_0^{\pi/2} d/2 \cdot \sin\alpha \cdot d\alpha}{\pi/2} = \frac{d}{\pi} \tag{4}$$

But this would neglect the fact that the synapses whose centres lie within the boundaries thus defined appear in the section with a probability dependent on their orientation: the ones that lie parallel to it may not touch it at all. Taking this into account, Mayhew (1979) computed the virtual increase of the thickness of the section as $d\pi/4$ (d is the diameter of the synaptic junctions) and the formula relating the real density of synapses N/mm^3 to the number counted in a section of area F and thickness s thus becomes

$$\frac{N}{\text{mm}^3} = \frac{N_c}{F(s + d\pi/4)} \tag{5}$$

The varying angle formed by the synapses with the image plane must also be considered when assessing the diameter d of the synapses from the average size of their projection δ. In analogy to the case of spherical bodies, we get for the synapses

$$d = \frac{2}{\pi}\left(\bar{\delta} - \frac{4s}{\pi} + \sqrt{4s\bar{\delta} + \left[\bar{\delta} - \frac{4s}{\pi}\right]^2}\right) \tag{6}$$

(Schüz 1981b, with improvements suggested by Palm).

Again, with synapses of different size a further improvement has to be made. Supposing the variance v as known, Schüz and Palm (1989) used the formula:

$$d = \frac{2}{\pi}\left(\bar{\delta} - \frac{4s}{\pi} + \sqrt{4s\bar{\delta} + \left[\bar{\delta} - \frac{4s}{\pi}\right]^2 - v}\right). \tag{7}$$

Formula 3 for the neurons of varying sizes was similarly amended by the same authors:

$$d = \frac{2}{\pi}(\bar{\delta} - s) + \sqrt{\left[\frac{2}{\pi}(\bar{\delta} - s)\right]^2 + \frac{4\bar{\delta}s}{\pi} - v}\,. \tag{8}$$

The formulas (7), (8) which consider the variance explicitly give a smaller value, for the size of synapses and of neurons respectively, than the formulas (3), (6). The two sets of formulas may be taken as upper and lower bound estimates.

Chapter 4

Density of Neurons

Contrary to the density of synapses which, as we shall see, does not vary much from animal to animal and from place to place in the same brain, the number of neurons per unit volume varies a great deal. As a rule the density of neurons is lower in larger brains, a fact which is explained in a qualitative way by the larger size of the neurons in the larger animal (see chapter 34).

The number of neurons in 1 mm^3 of mouse cortex was estimated, by rather crude methods, as 2×10^5 in our earlier papers (Braitenberg, 1978b). Since this number plays a crucial role in some of our quantitative arguments, Schüz and Palm (1989) set out to obtain as precise as possible a figure with the best available methods. Their study includes a comparison of different areas of the cortex in a number of mice.

The cortical areas examined were area 8, area 6 and area 17 of the Caviness map (Fig. 74). The material was taken from three white laboratory mice, female and between 4 and 6 months old, after transcardial perfusion with 0.9% NaCl followed by fixation in 1% paraformaldehyde and 1% glutaraldehyde in phosphate buffer. The preparation of the tissue was essentially the same as that recommended by Palay and Chan-Palay (1974) for electron microscopy, including staining with osmic acid (2% OsO_4 and 7% dextrose in phosphate buffer) and embedding in Epon-Araldite. Counting neurons on the semithin sections from the same blocks that could later be used to measure the density of synapses on electron micrographs seemed the best way to avoid the uncertainties which arise in the quantitative comparisons between tissue elements subjected to different methods of fixation and embedding.

The semithin sections were between 1 and 4 μm thick. Care was taken to cut the cortex in the "vertical" direction, perpendicular to the surface. The sections were stained with Azur II and Methylene blue. In material treated with Osmium, this stain seems to darken mostly the membranes and to a lesser extent other cellular components. Neural cell bodies cannot be as easily recognized in such preparations as in Nissl preparations, but the nuclei of the nerve cells stand out very clearly, can be recognized unambiguously and, what is more, their size can be easily measured.

The thickness of the sections was determined with the microdrive of a microscope using a 100× objective. The distance between the lowest and the highest objects that could be focussed in the section was measured. The depth of focus of a large aperture 100× lens is well below one micrometer, allowing a fairly accurate setting of the plane of focus. The method was tested on one block of tissue the thickness of which was first measured macroscopically and then cut into a complete series of semithin sections. The number of sections multiplied by their thickness measured by means of the microdrive of the microscope corresponded very well to the macroscopic measurement (a block of 92 μm height gave 47 sections, whose measured thickness was 2 μm).

The outlines of the nuclei of nerve cells were drawn with a camera lucida. Also drawn were nuclei of glia cells and blood vessels, for reasons which will become apparent later. Each sample area measured was between 13000 and 40000 μm^2. For each of the three cortical areas examined in each of the three mice, ten such sample areas were measured, two in each of the cortical layers II to VI.

We mention these technical matters because they may be at the base of the discrepancy between the counts by Schüz and Palm (1989) and those of other authors.

The results from our laboratory are summarized in the following diagrams. Figure 9 shows for the entire sample the average diameter of the neuronal nuclei according to areas and layers. These nuclei are of remarkably constant size, except for the nuclei in area 8, which are consistently between 10 and 20% larger than the others. The size of the nuclei had to be measured, as you may recall, in order to apply the Abercrombie correction to the counts. Figure 10 presents the result of the counts, corrected and expressed as number of neurons per volume, for the five layers and three areas, again averaged over the three mice. Layers II, IV

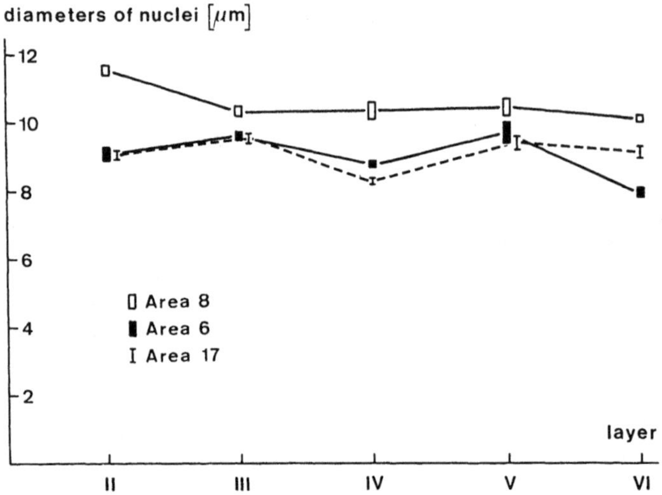

Fig. 9. Size of the nuclei of cortical neurons in three different areas and five different layers. Averaged data from three mice. The *size of the symbols* indicates the difference between two different stereological methods used in Schüz and Palm (1989)

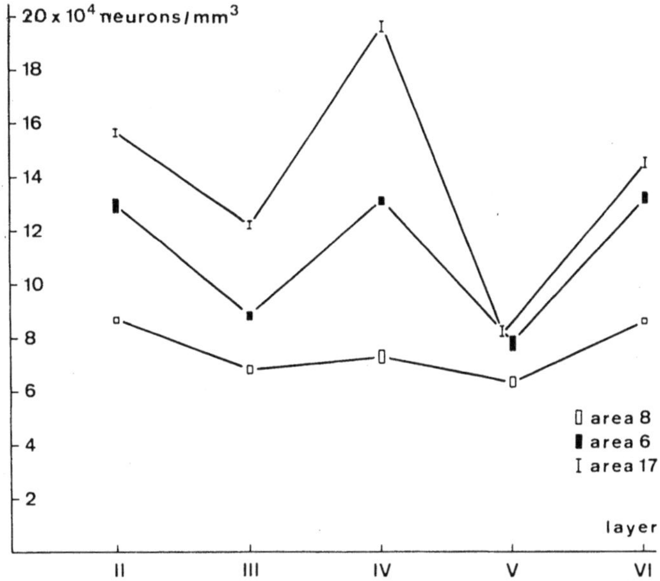

Fig. 10. Density of the neuronal population in three different areas and five different layers. Area 17 has most neurons per unit volume, especially in the upper layers, and the frontal area 8 the fewest. The *upper and lower edges of the symbols* indicate the upper and lower limits, obtained by two different stereological approaches. (Schüz and Palm, 1989)

and VI are consistently richer in neural cell bodies than the other layers, less clearly so in area 8. Figure 11 displays the same data in a different way, to emphasize the different densities in the three areas, area 17 being the richest and area 8 the poorest.

The density of neurons in the cortex, averaged over the entire sample (three mice, three areas, five layers) in the Schüz and Palm counts is $9.2 \times 10^4/\text{mm}^3$. This is about one half the value assumed in our previous reports (e.g. Braitenberg 1978b). It is also less than the values published (for the mouse) by Tower and Elliott (1952), Bok (1959), Sholl (1959) and Heumann et al. (1977), which are all between 1.4×10^5 and $2 \times 10^5/\text{mm}^3$. Since the only older report which comes up with figures as low as those of Schüz and Palm is that by Cragg (1967), and since this author used frozen sections, we assume that all the other counts were affected by the shrinkage that occurs with paraffin or celloidin embedding.

An interesting invariant of cortical structure has been proposed by Bok (1959) and again by Rockel et al. (1980). These authors claim that the number of neurons below a given area of cortical surface is remarkably constant for different areas and different mammalian species (see also chapter 34). Figure 12 displays our counts as number of neurons per area of cortex (rather than volume). The figure of about 10^5 per square

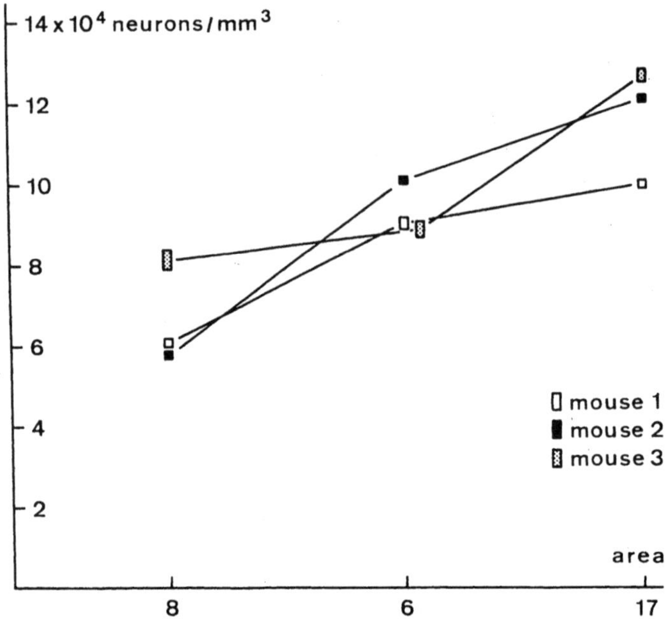

Fig. 11. Density of neurons in the three areas, averaged over all layers. The density in the occipital region is almost twice as high as in the frontal region. *Size of the symbols:* meaning as in Figs. 9 and 10. (Schüz and Palm, 1989)

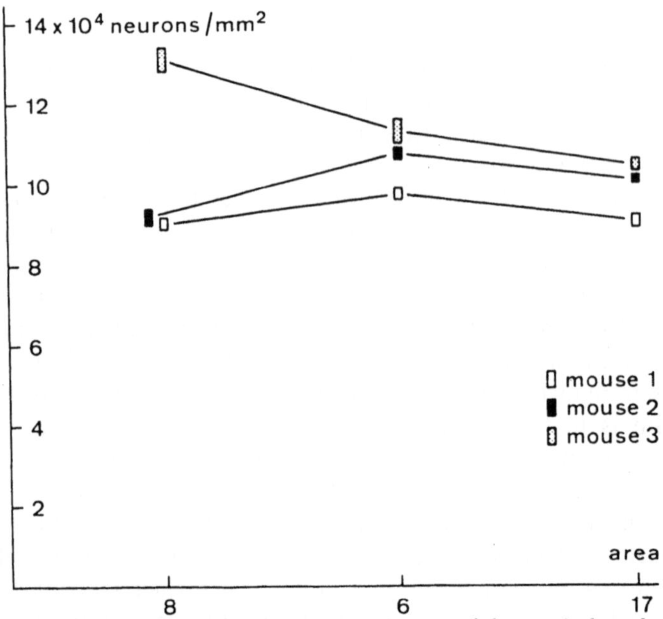

Fig. 12. The number of neurons per unit area of the cortical *surface* remains fairly constant in all three areas. Again the *symbols* indicate upper and lower limits. (Schüz and Palm, 1989)

millimetre corresponds to that given by Rockel et al. (1980). We are as yet unable to attach a meaning to this remarkable constancy.

The total number of neurons in the mouse cortex can be estimated from their number in a unit volume. The volume of the cortex, including the hippocampus, amounts to about 87 mm^3 in one hemisphere. With a density of $9{,}2\times10^4/mm^3$, the total number of cortical neurons in the mouse is 1.6×10^7 in both hemispheres taken together. We remind the reader that the estimates for the human cortex hover around 10^{10} (e.g. Haug, 1986; Braendgaard et al., 1990), almost 1000 times more neurons than in the mouse.

If only the volume of the neocortex is considered which is 56 mm^3 in one hemisphere, the total number of neurons in the mouse is 1×10^7 in both hemispheres.

Chapter 5

Density of Synapses

Only the electron microscope provides the resolution which makes it possible to recognize individual synapses reliably.

Again, we can base ourselves on the paper by Schüz and Palm (1989) which made a point out of counting neurons and synapses on the same material in order to facilitate comparison.

The blocks of tissue used were the same as described in the previous chapters. After the "semithin" sections were cut from each block for light microscopy, the adjacent sections were cut with the ultramicrotome. This made it easy to identify the cortical layers in the electron micrographs by marking them on the low power light micrographs and transferring the coordinates to the sections prepared for electron microscopy. The "ultrathin" sections chosen for the counts of synapses were about 70 nm thick as judged from their interference colour. (This can be done fairly accurately by means of a standard colour table). The sections were stained with uranyl acetate and lead citrate. Electron micrographs were taken at a magnification of ×28 000. Samples were taken from Layer I, Layer III, IV, V and VI. Each sample was represented by a montage of between 8 and 14 adjacent micrographs covering a total area of 230 to 320 μm^2. Since it is known from Wolff (1976) that the synaptic density in the cortex varies with the distance from larger blood vessels, being minimal close to the veins, care was taken to select samples sufficiently far away from such vessels.

For the purposes of our counts a synapse was defined pragmatically, without regard to the problems of interpretation which arise when the cytological details are related to their function. A synapse was counted when both the presynaptic and the postsynaptic thickening were recognizable and when there were at least three vesicles on the presynaptic side not farther away than 0.1 μm. Synapses were counted irrespective of their orientation in the section, including the ones cut tangentially, provided the above-mentioned conditions were met. Thickenings of two pieces of cell membrane in close apposition without the presence of vesicles were occasionally observed and recorded, but not included in the counts.

The size of the synaptic junctions (i.e. the diameter of the postsynaptic disc) was measured as indicated in Fig. 13. Their "real" size was computed from the distribution of their sizes on the electron micrographs by means of the formulas (6) and (7), Chapter 3. The thickness of the sec-

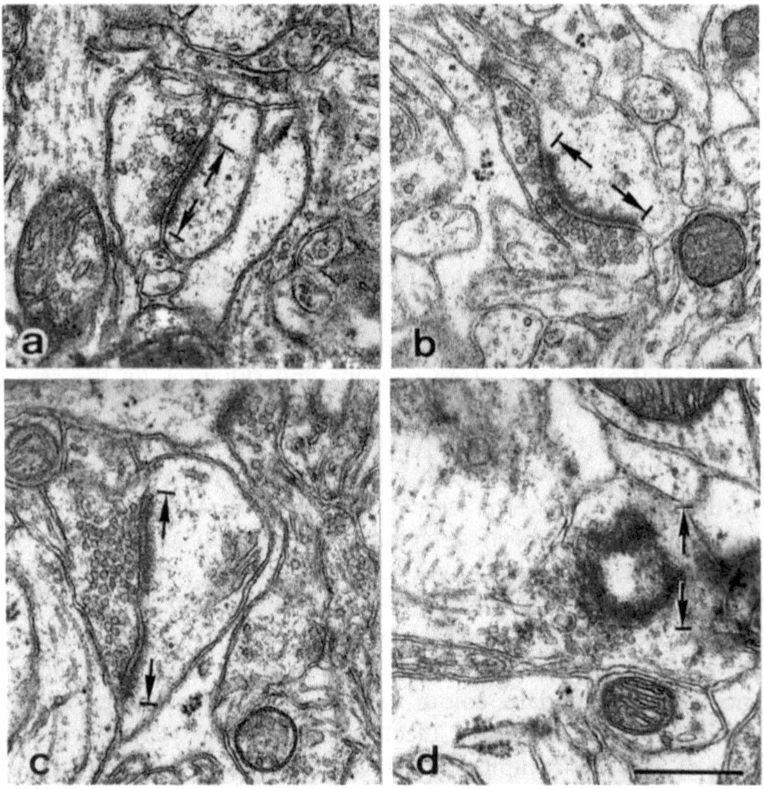

Fig. 13 a–d. Some examples of synapses on electron micrographs, to show how their size was measured between the lines marked by arrows. Synapse **a** is straight, **b** is curved, **c** has an interrupted postsynaptic thickening and **d** is cut tangentially. The *bar* at **d** is 0.5 μm long. (Schüz and Palm, 1989)

tion which enters in these formulas was taken as 50 nm, rather than 70 nm, taking into account the fact that synapses can only be recognized when a sufficiently large segment is contained in the section.

Figure 14 shows the results. The size of the synaptic junctions does not seem to vary much between different layers and areas, the values for the different samples ranging from 320 to 380 nm. The only possible correlation with the depth in the cortex is in Area 6, where the synapses seem to be a little smaller in layers IV and V than in the other layers.

Similarly, the number of synapses between layers I and IV stays around the value of $9\times10^8/mm^3$, only to fall off slightly in layer V and especially below, where it reaches a density of about $7\times10^8/mm^3$ (Fig. 15).

The grand average of synaptic density for the entire sample (corrected for space taken up by cell bodies and blood vessels as will be described in Chap. 6) is $7.2\times10^8/mm^3$. This is slightly less than the density which

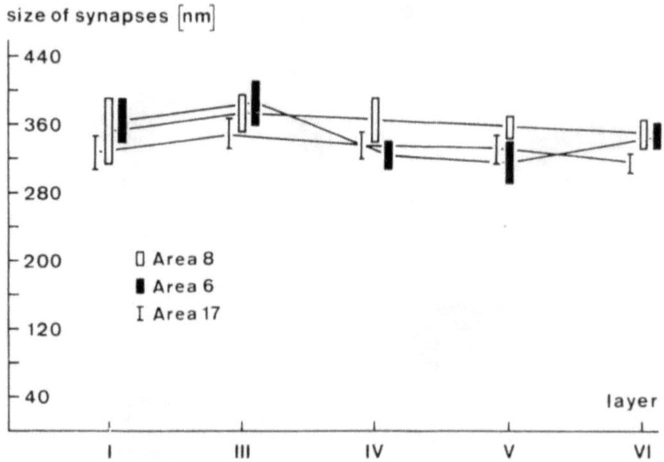

Fig. 14. The size of synapse: the three areas is remarkably constant. The *size of the sy* ence between Eqs. (6) and (7), Chapter 3

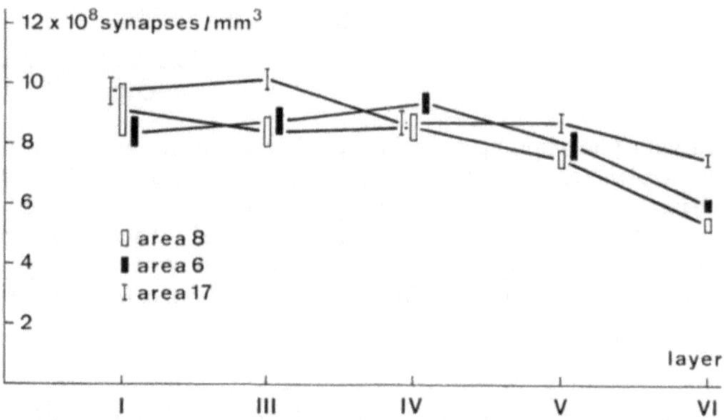

Fig. 15. Density of synapses in different layers of the three areas. The *size of the symbols* indicates the difference between upper and lower limits. (Schüz and Palm, 1989)

corresponds to a uniform distribution of the synapses spaced at 1 μm from each other in every direction.

5.1 Light Microscopy of Synapses

Synapses are generally considered to be invisible in the light microscope because of their small size. This is indeed true as far as transmission light microscopy is concerned (see, however, Chaps. 9, 10 and 22). With dark field illumination the location, if not the shape, of particles far smaller than the wavelength of the light can be seen, provided their opti-

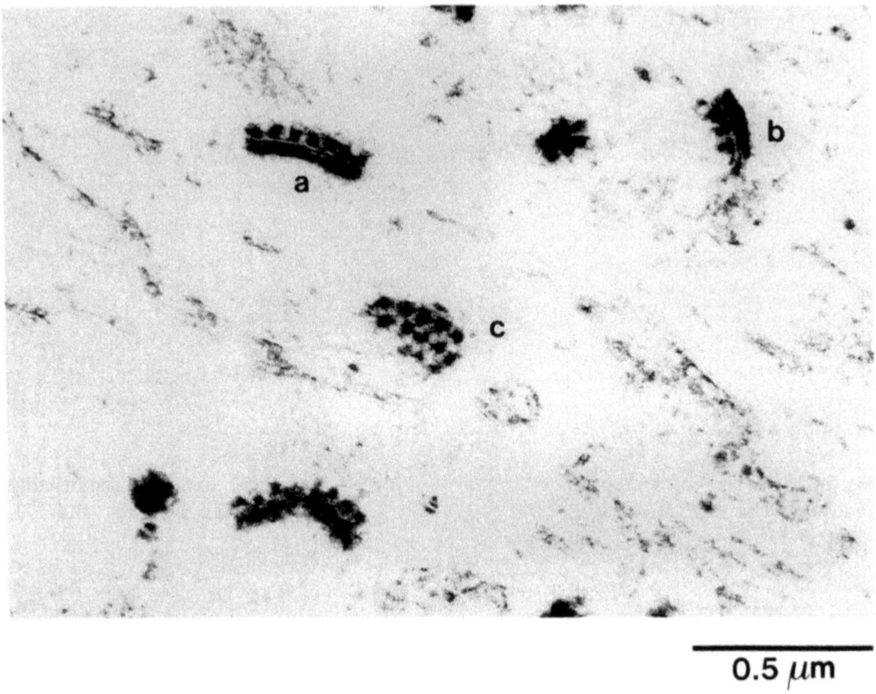

Fig. 16. Synapses stained with phosphotungstic acid according to Bloom and Aghajanian (1968). In *a* and *b* the plane of the section is fairly perpendicular to the plane of the synapse. The uninterrupted dark material is on the postsynaptic side, the one consisting of separate lumps on the presynaptic side ("dense projections"). There is a thin dark line in between, located in the synaptic cleft. In *c* the arrangement of the presynaptic dense projections in a regular hexagonal grid can be seen, the synapse lying parallel to the plane of the section

cal properties (absorption and/or refractive index) are sufficiently different from those of the background.

The stain with phosphotungstic acid introduced by Bloom and Aghajanian (1966, 1968) darkens the pre- and postsynaptic elements of synapses with astonishing selectivity (Fig. 16). M. Dortenmann, of our laboratory, found that the selectivity can be increased by suppressing the stain of mitochondria which also tend to accumulate the tungsten. This can be done by low pH fixation and/or by exposure to ultrasound before applying the stain.

The background can be kept so clear that even sections 300 to 500 nm thick can be used in the electron microscope. The very same sections, photographed with a high power lens in dark field lightmicroscopy (Fig. 17), provide an impressive display of the density of the synapses in the tissue (Braitenberg, 1981; Braitenberg and Schüz, 1983). The comparison of the same region of the section in the two kinds of pictures (Fig. 18) shows that most of the light speckles in the dark field micro-

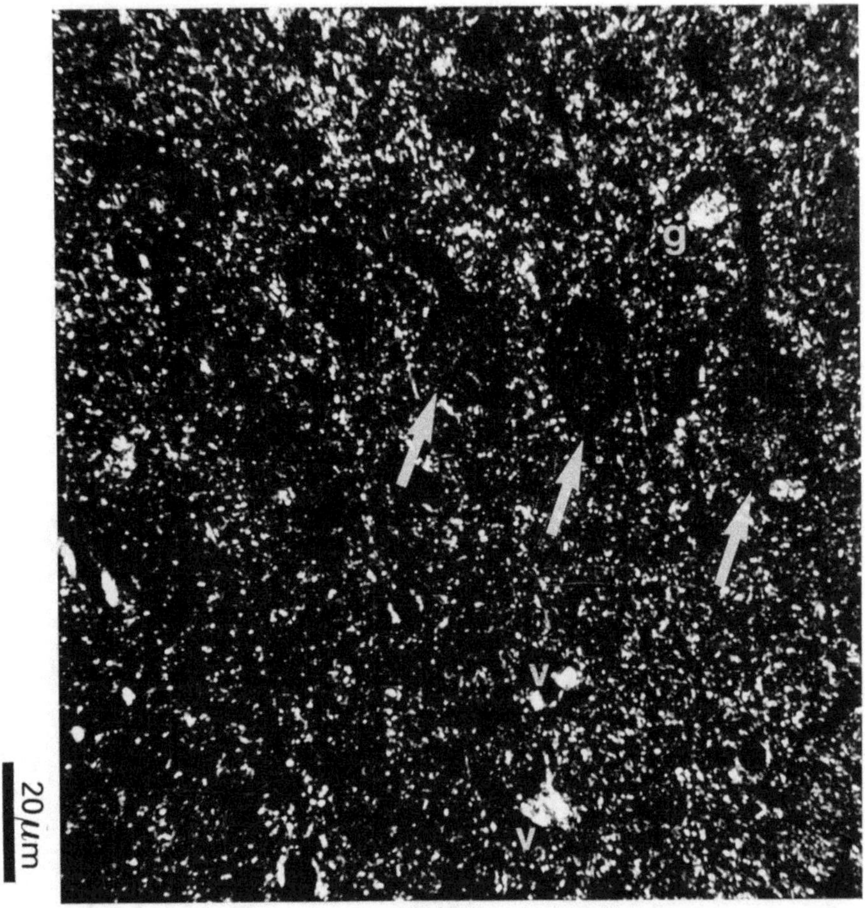

Fig. 17. Dark field photograph of synapses *(bright spots)* in the third layer of the mouse cortex. The stain is phosphotungstic acid, of the same quality as in Fig. 16. The stain is not absolutely specific for synapses: some elements associated with blood vessels *(v)* and some material in the nuclei of glia cells *(g)* also lights up in the dark field micrograph. The nerve cell bodies appear as black zones *(arrows)* surrounding the nuclei. Some apical dendrites can also be recognized. (Braitenberg, 1981)

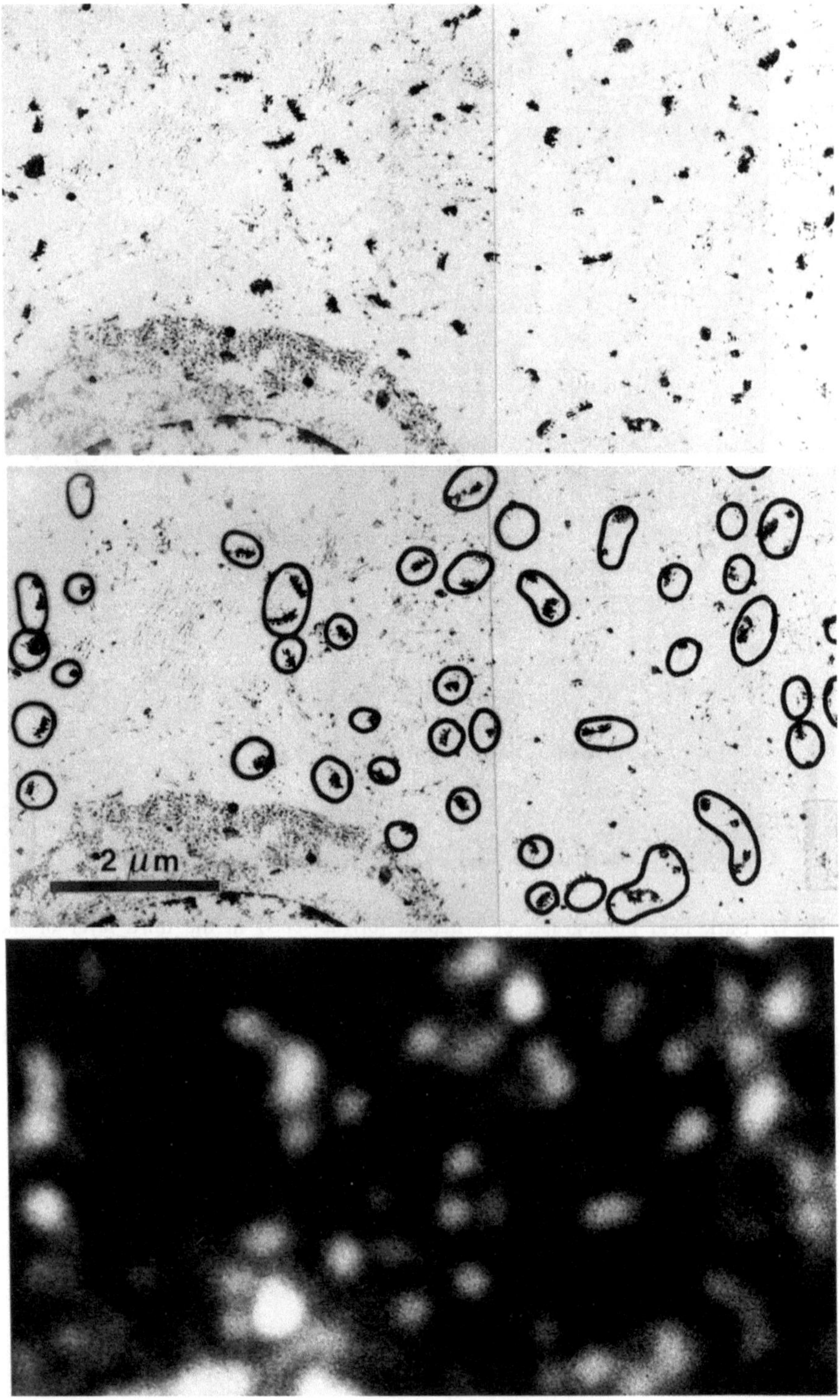

Fig. 18

graph do indeed correspond to the location of synapses. Unfortunately, for reasons of wave optics, two or more synapses that lie too close together appear as one point in the dark field picture. For this reason this method cannot really be used for quantitative neuroanatomy.

Fig. 18. Correspondence between an electron micrograph *(above)* and a dark field light micrograph *(below)* of the same section stained with phosphotungstic acid. If outlines of the light spots of the dark field picture are superimposed on the electron micrograph *(middle)*, most of them are seen to correspond to synapses, sometimes to more than one synapse

Chapter 6

Comparison Between Synaptic and Neuronal Density

The average number of synapses per neuron is a value which gains importance in theoretical considerations about cortical function. It reveals the maximum possible degree of dispersion of the signals emanating from one neuron, or conversely, the maximal possible convergence of signals from different sources onto one neuron. On the other hand, convergence/divergence may not be the essence of the story. If we suppose that one neuron may be connected to another not just by one but by any number of synapses, the numerical ratio of synapses and neurons in the tissue is an indication of how fine the tuning of the influence can be. Our personal preference for one or the other interpretation of the number of synapses per neuron will be apparent in a later section of this book.

In order to obtain this ratio, we had to apply a correction to our measurements of the density of synapses. Although both counts, that of neurons and that of synapses, were obtained on sections from the same block of tissue, the sample for the synapses was biased in a way that that of neurons was not. The density of the synapses was measured in the neuropil, i.e. in cortical tissue not containing cell bodies or blood vessels. On the other hand, the number of neurons was put in relation with the volume of the entire cortex. What proportion of the cortical volume belongs to the neuropil? This was measured in the light microscope on semithin sections. Figure 19 shows for the different cortical layers, the proportion of the area of neuropil, i.e. the area of the section not showing neural or glial cell bodies, over the entire area. Except for layer I, which is almost exclusively neuropil, the values are between 80 and 90%. In this the blood vessels have not yet been considered. Their part of the volume was determined on micrographs at lower magnification and was found to be 4%. According to a well-known theorem of stereology (Underwood, 1970), in a volume consisting of several distinct fractions, the relative proportions of these fractions appearing on sections are the same as those in the volume. Therefore our measurements of the fraction of micrograph area representing the neuropil translate into the fraction of cortical volume occupied by neuropil. It amounts to 84%. The number of synapses, in order to be compared to the number of neurons in the same volume of cortex, must be corrected by a factor 84/100.

This done, the ratio of synapses and neurons in the mouse cortex, averaged over the entire sample, can be computed as

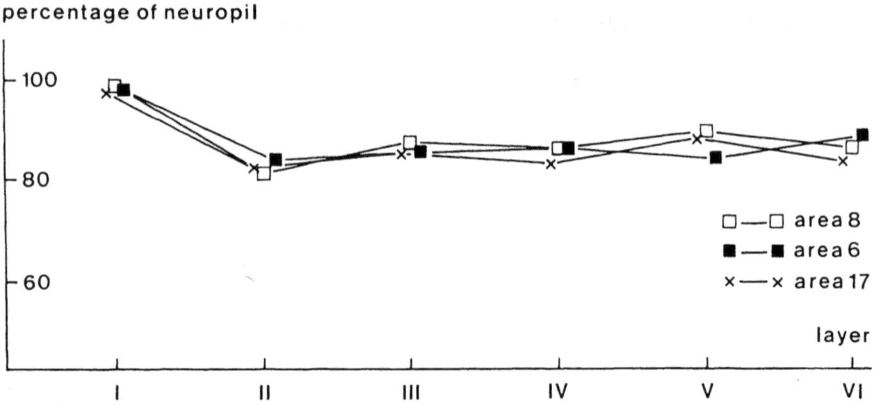

Fig. 19. Percentage of "neuropil", i.e. everything minus nerve cell bodies and nuclei of glia cells, in six different layers and three different areas. The percentage was calculated in terms of surface area on microphotographs, and should be approximately the same as the percentage in terms of volume

$$\frac{7.2 \times 10^8}{9.2 \times 10^4} = 7826 : 1 .$$

This is in good agreement with the figure given by Cragg (1967) for the mouse. In different animals different ratios have been reported. The values for area 17 vary between 2300 synapses per neuron (area 17 of the monkey) and 17000 synapses per neuron (area 17 of the rabbit). For review see Peters (1987). A ratio of 60000 synapses per neuron was reported for the monkey motor cortex (Cragg 1967). We will take up comparative aspects again in chapter 34.

The total number of synapses in the mouse cortex (including the hippocampus) turns out to be of the order of 10^{11} (an estimated 10^{14} in the human cortex).

Chapter 7

Density of Axons

The sum total of the length of all the nerve cell processes (dendrites and axons) in a given volume of brain substance provides, in a sense, a measure of the information-handling capacity of the nerve tissue. From a different point of view, we may take it as indicating the complexity of the neural operations performed there.

We chose to call this measure density of fibres, because we intend to use it in quantitative comparisons with other densities: of neurons, of synapses, of dendritic spines. In reality, its dimensions are not those usually associated with the concept of density (mass per volume, number per volume or volume per volume), but length per volume, or length to the power of –2. One may think of this as the inverse of the cross-sectional area of the elements contained in the tissue, or as the number of cross-sections per unit area, and this is indeed one of the ways we measured the density of axons and dendrites on electron micrographs. It should be noted that the density of fibres is not necessarily a measure of the degree of internal connectedness of the nerve tissue. In fact, we may imagine two kinds of tissue, one containing a large number of short fibres and another a smaller number of longer fibres, both yielding the same density of fibres. However, it would seem that the degree of internal connectedness is larger in the tissue with the longer fibres.

Of the three kinds of fibres which make up the neuropil, we disregard the fibres of the glia which play no further role in our theoretical reasoning. On the other hand, the density of both sorts of neuronal cell processes, axons and dendrites, is of prime importance, since it enters into various calculations of the probability of the connections between neurons.

We followed three different approaches to estimate the density of axons in the cortex.

The first (Braitenberg, 1978b) was based on Golgi pictures of neurons. Since the results of our scrutiny of Golgi preparations will not be presented until a later chapter, we shall simply discuss the principle here and leave the details until then. The length of all the axonal branches of one neuron can be easily measured in the preparations, provided the neuron is sufficiently isolated (i.e. the Golgi stain has not darkened too many neurons in the neighbourhood). There are two catches, however. If the picture of the neuron is projected onto a plane, as it necessarily is when it is drawn with the camera lucida, information about the third dimension is lost. This information can be obtained if the movement of the microdrive of the microscope is recorded during the drawing procedure. Simple geometry will then suffice to calculate the true length of the fibre from the length of its projection. We did not have the equipment to do this in a routine way, as other laboratories have, and also thought it unnecessary to measure the length of every branch if all we were after was the sum total of their lengths. This we intended to obtain from the sum of the lengths of their projections by supposing that the angles they formed with the plane of the section were distributed randomly.

The second catch is that only very rarely is the full axonal arbor of one neuron contained in a single section. It is notoriously difficult to follow axonal branches into the neighbouring sections, for which reason we had to rely also in this case on a statistical correcting factor (for details see Chap. 16).

If we distinguish a number of distinct neuron types, we may measure the axonal length for each type separately. Knowing (or estimating) the relative frequency of each type, we can calculate the total axonal density in the tissue.

In our first effort (Braitenberg, 1978b) this took the form of a rough estimate, 1 or 2 km of axons per mm^3.

The other method we used to assess the total axonal length in a unit volume ("axonal density") was based on electron micrographs. On these it is possible to recognize the fragments of axons among those of dendrites, glia and other tissue components. Because of the fundamental result of stereology already mentioned (Underwood, 1970), if we measure the areas of the sections of all the axonal fragments, the sum of these areas is to the total area as the volume of the axons in the tissue is to the total volume. Having thus determined the percentage of the cortical volume occupied by axons, it is sufficient to divide this volume through the average cross-sectional area of the axons to obtain their total length. We are aware that the average of the inverse of a certain set of measurements is not the inverse of the average, but are not troubled by this because of the small variation of axonal cross-sections in the mouse cortex. The cross-sectional area can be easily determined by taking the smallest diameter of the outlines of the axonal fragments, which are in general elliptic in shape, and by assuming that these are in reality more or less oblique sections through cylindrical fibres.

The measurements (Schüz, 1989) on samples of a size varying between 38 and 48 μm^2 yielded the following percentages for the area occupied by the different tissue components:

Axons	34%
Dendrites	35%
Spines (necks and heads)	14%
Glia	11%
Extracellular space	6%

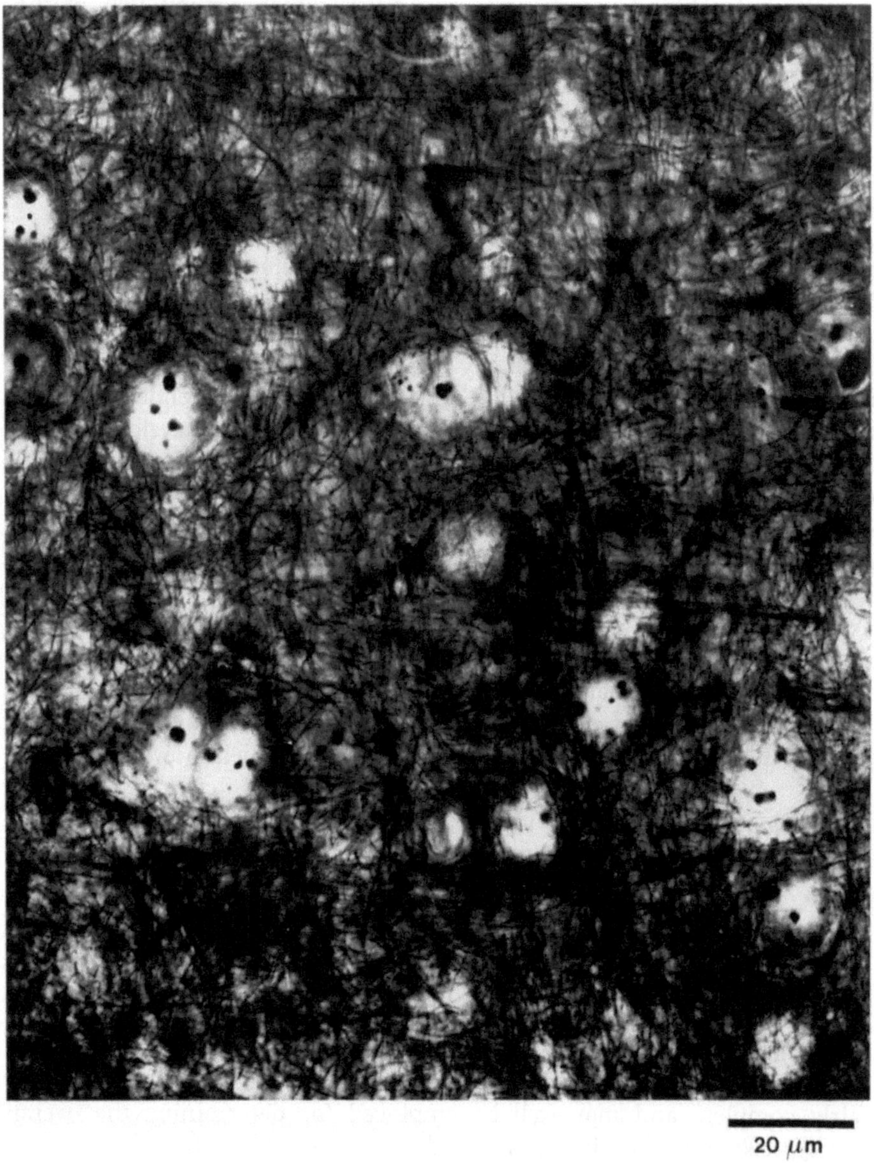

Fig. 20. Reduced silver stain after Bielschowsky, modified by V. Staiger. Layer V of the mouse cortex. The staining of axons is fairly complete, while the staining of dendrites and neural cell bodies is suppressed, except for some lumps inside the nuclei. The density of axons on such preparations is higher than 1 km/mm^3. Note that the enlargement is the same as in Fig. 17. (Braitenberg, 1981)

Correcting for blood vessels and cell bodies, which occupy 16% of the cortical tissue but were not contained in the samples, this becomes for the axonal components 29% of the cortical volume, and for dendrites about the same. These values are in good agreement with those given by Foh et al. (1973) for the cat cortex.

The average axonal diameter measured on our electron micrographs was slightly below 0.3 μm. From this we can calculate that the 29% of cortical volume occupied by axons corresponds to an axonal density of 4.1 km per mm^3. This is close enough to the value determined on the Golgi preparations, but does not quite invalidate the suspicion that Golgi staining leaves much of the axonal tree unstained.

Finally, we measured the density of the axonal fibre population on reduced silver preparations (Fig. 20). This was originally intended to act as a check on the selectivity and completeness of a stain developed by V. Staiger in our laboratory, a modification of the old Bielschowsky procedure requiring impregnation with silver nitrate first and with a complex silver compound afterwards, which is then reduced by means of photographic developer. The decisive modification which suppressed the staining of cell bodies and dendrites completely while producing a very dense axonal network, was a prolonged treatment of the pieces of tissue in glucose solution before sectioning and staining. Fixation in a mixture of glutaraldehyde and formaldehyde administered by perfusion (the optimal fixation developed by electron microscopists) was certainly another reason for the drastic improvement upon the original Bielschowsky preparations.

(Incidentally, the Staiger procedure produces a very distinct stain of the "parallel fibres" in the cerebellum which are, perhaps because of their small caliber, notoriously resistant to other types of reduced silver stain).

On Staiger preparations the distance between neighbouring fibres taking a roughly parallel course was determined with the light microscope and found to be about 1 μm. Schematizing the distribution of the fibres as an orderly network with fibres running in the three orthogonal directions, this corresponds to a density of 3 km per mm^3. Evidently the reduced silver stain (when it succeeds, which is not always the case) is fairly complete and may well be employed for determining the overall characteristics of the distribution and orientation of the fibres.

Chapter 8

Comparison Between the Densities of Neurons, Synapses and Axons

The fact that 4 km of axons belonging to 9.2×10^4 neurons allot more than 4 cm of axon to each neuron is surprising, especially since direct measurements on Golgi preparations have led to values less than half that figure. It must be noted, however, that the measurements in Braitenberg 1978b which refer to pyramidal cells (the classification of cell types will be introduced in Chap. 15) do not consider the re-entrant axons of these cells. These may make considerable terminal arborizations which are only rarely stained by the Golgi procedure and if they are, are practically never seen in continuity with the other part of the axonal tree.

An alternative explanation for the high density of axons in the cortex is that much of this axonal population is of external origin, being derived from axons reaching the cortex from other regions of the nervous system. If this were so, the lower value for the axonal density which we obtain by multiplying the number of cortical neurons by the length of their axons in Golgi preparations would not be surprising. It would actually make it possible to estimate the proportion of axons of extra and intracortical origin, a value which is difficult to obtain by other means. In our opinion the extracortical afferents (specific sensory, thalamic and brain stem afferents) contribute, all told, little to the cortical network, but proof of this is lacking. We know that in primary sensory areas, at the cortical level where the input fibres terminate, up to 20% of the synapses are supplied by extracortical afferents (White, 1986) but in other areas and other layers the percentage is likely to be much lower.

Be that as it may, the comparison of axonal density with synaptic density speaks rather in favour of the higher value for the axonal length. If 7.2×10^8 synapses are served by 4 km of axons, there are on average 180 synapses on 1 mm of axon, or one synapse to every 5.5 μm. If the axonal density is only 1 km/mm^3, the synapses would be crowded at an average distance of 1.38 μm along the axon. Since there are undoubtedly stretches of axons which do not form any synapses, for instance where the axon is myelinated, with the average spacing of 1.38 μm, we would have to assume that in many places the synapses are spaced at even shorter distances along the axon.

The question of the spacing of synapses along the axon is an important one, especially in connection with the problem of the "specificity" of neuronal connections in the cortex. If one pictures intracortical axons as finding their way to their target neurons and making all the synapses there, one would expect long stretches of axon without synapses and a very high density of synapses at the place of termination. With an average spacing of 1.38 to 5.5 μm, which results from our calculation on the basis of 1 to 4 km of axons per mm^3, the crowding of synapses at the terminal sites would actually be impossibly dense. This is one of the strong arguments against the supposition of a high specificity of the connections in the cortical network.

Chapter 9

Microscopic Evidence of the Spacing of Synapses on Axons

The question of "specificity" being of prime importance for our general view of the cortical network, we decided to tackle the problem directly by counts of synapses formed by selected axons. Our first approach was with the electron microscope (Schüz and Münster, 1985), a task requiring rather tedious work on serial sections. We then discovered a short cut, permitting us to count synapses on much larger samples, and much more easily, by means of the light microscope (Hellwig 1992; Hellwig et al., 1994).

On sections prepared for electron microscopy and stained in the usual way with heavy metals (Osmium, Uranium) it is very difficult to follow individual axons through serial sections. Moreover, it is wellnigh impossible to identify the cells of origin of the axons on such preparations. We decided therefore to use the combination of the Golgi technique and electron microscopy which had been developed by several authors (Blackstad, 1965, 1981; LeVay, 1973; Fairén et al., 1977). The method is tricky and requires a compromise between a sub-optimal Golgi stain and sub-optimal electron microscopy. For one thing, the Golgi stain works really well only if the tissue is fixed directly in the dichromate solution which the Golgi method requires, without previous aldehyde fixation. On the other hand, the best results in electron microscopy are obtained after perfusion of the live tissue with aldehydes. A further difficulty resides in the coarse precipitation of silver chromate, which fills the neurons once they have been successfully stained in the Golgi preparations. If such preparations are later reimbedded in plastic in order to be cut into very thin sections by the ultramicrotome, the granular precipitate is often torn out of the cell by the knife and destroys the tissue. Various technical solutions have been proposed for this problem, of which we found Blackstad's photochemical method the most satisfactory for our purpose. The cells which accept the Golgi stain are treated with light, which reduces the precipitate to metallic silver in the form of fine grains. These are no obstacle to ultrathin sectioning, but provide a reliable marking of every part of the Golgi-stained cell on the electron micrographs. In addition to the Osmium-Uranium stain, we treated the sections with phosphotungstic acid according to Bloom and Aghajanian (1966), thereby enhancing the stain of the pre- and postsynaptic thickenings.

Four horizontal collaterals of a descending axon of a pyramidal cell (see Chap. 15 for the classification of neuronal types) were chosen for the analysis in Schüz and Münster (1985). The reason for this choice was the belief that most of the synapses in the cortex have axon collaterals of pyramidal cells as their presynaptic element. The Golgi-stained cell was photographed and drawn with the camera lucida (Fig. 21, above); care was taken to capture every kink of the collaterals in order to facilitate the comparison of the electron micrographs with the light microscopic picture.

On the serial electron micrographs which were then made, the axon of the stained cell with its synapses was easily recognized. The main (descending) axon was followed for a length of 193 μm on serial sections, and a total of 169 μm of its collaterals were also surveyed for synapses. The initial segment of the axon carried 12 synapses in which it functioned as the *postsynaptic* element. The rest of the descending axon had hardly any synapses and the same applied to the initial segment of the collaterals after the branching point. The synapses were present on the more distant portions of the collaterals: 17 synapses on 60 μm in the case of the collateral, which was followed for the longest distance (Fig. 21 below). A total of 28 synapses was counted on the 332 μm of axon which were examined, corresponding to the density of 1 synapse every 12 μm. This value was much lower than expected, but probably does not represent the true average over the whole axonal tree. Indeed, extrapolating from what we found in the portion of the axon examined here, the collaterals could be expected to carry the majority of their synapses (about one every 3.5 μm) on their more distal segments, which make up most of the total axonal length belonging to the neuron. The proximal segments, which were free of synapses, are only a very small part of the entire tree, so that the grand average could be close to one synapse every 3 to 4 μm.

The most interesting finding in this tedious analysis was the one which actually discouraged us from carrying it any further. It became clear in the reconstruction of the axon collaterals from serial electron micrographs that most of the synapses were located on swellings of the axon which had previously been recognized on the Golgi picture and which were also quite apparent on the electron micrographs. (Evidence for this had been presented for a number of non-pyramidal cells by Peters and Proskauer 1980; Somogyi and Cowey 1981; and Somogyi et al. 1983). On the basis of this observation it seemed opportune to continue the study with the light microscope in order to lend it a broader statistical base.

If every bouton corresponded exactly to one synapse, the fact that all synapses reside on axonal swellings or "boutons" would make it possible to count synapses in Golgi preparations; but such is not the case. Already on the collateral shown in Fig. 21, there are two boutons carrying two synapses each. The first question (Hellwig, 1992) was then: how often does this occur? In an analysis of 38 boutons cut serially for the electron microscope, only two boutons were found to be presynaptic at two sy-

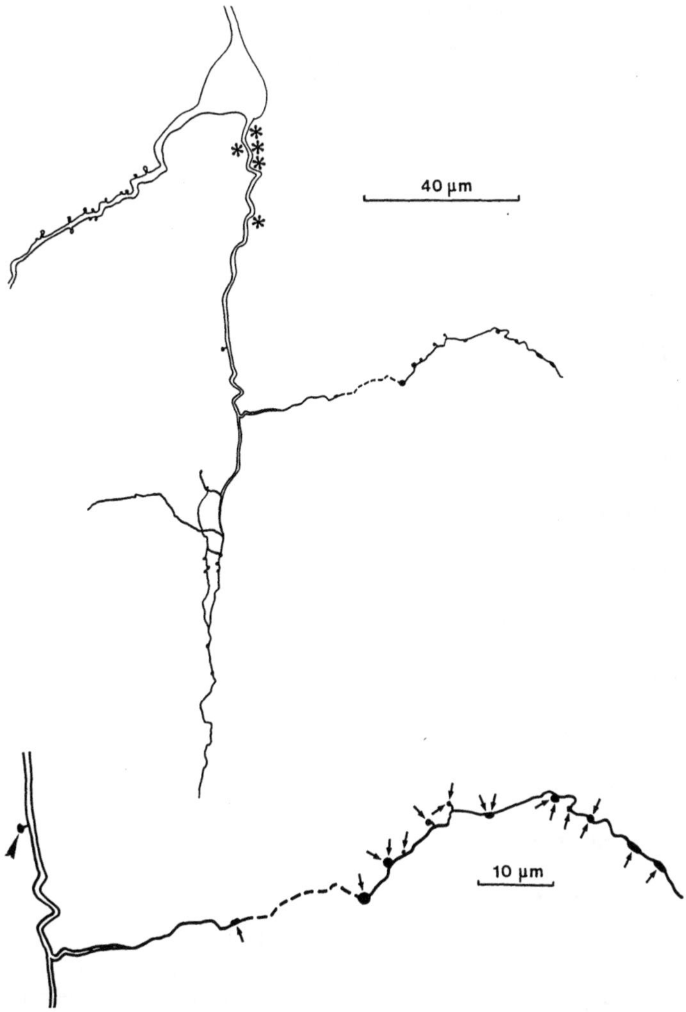

Fig. 21. Localization of synapses on the axon of a cortical pyramidal cell stained with the Golgi method and then examined on serial electron microscopic sections. The *asterisks* mark the position of synapses for which the initial segment of the axon was *post*synaptic. The synapses on the axon collateral (*arrows*, enlarged drawing below), all correspond to swellings which could already be seen in the Golgi preparation. The *arrowhead* on the left marks a synapse on a separate small appendix on the descending portion of the axon. (Schüz and Münster, 1985)

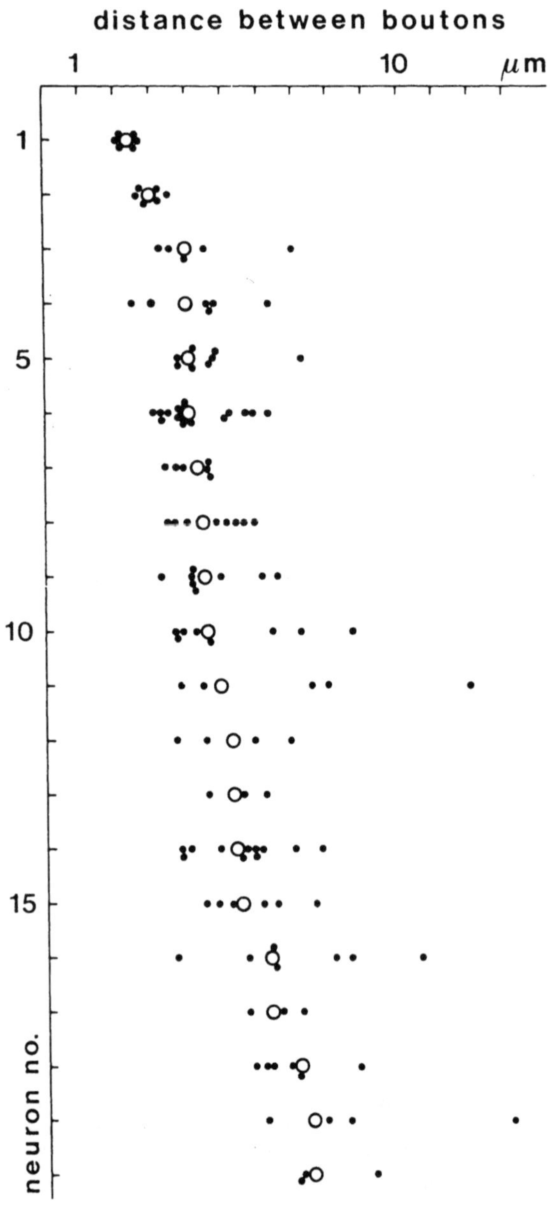

Fig. 22. Average distance between the boutons along the collaterals of 20 different pyramidal cells. The *empty circles* show the averages for all the collaterals of one neuron, the *black dots* are the averages for individual collaterals. (Hellwig, 1992)

napses and one at three. Thus counts of boutons in the light microscope can be expected to fall about 7.8% short of counts of synapses in the electron microscope.

The statistics of axon swellings was carried out on camera lucida drawings of 34 cortical neurons stained with the Golgi method. The neurons were of the pyramidal as well as the non-pyramidal type (see Chap. 15).

Care was taken to record every axonal swelling on the drawings. The length of the axonal segments was calculated from their length on the drawing and their extension in depth measured with the micro-drive of the microscope. This was important because the counts were ultimately to be expressed in terms of axonal swellings per axonal length, in order to relate them to the value which was obtained in the global comparison of the axonal and the synaptic density in the tissue (see Chap. 8).

The average distance separating boutons along the axons varied considerably from neuron to neuron and in some cases between different branches of the same axonal tree (Fig. 22), the most frequent value being one bouton every 4 to 5 μm. There is no great difference between pyramidal cells and non-pyramidal cells, the latter, especially the stellate cell variety, tending towards slightly higher values.

These counts are in remarkably good agreement with the values we had predicted on the basis of our measurements of synaptic and axonal density, under the supposition of an even distribution of the synapses along all the axons in the network without any special preference for "terminal" branches of the axonal tree. The counts speak in favour of the higher value for the axonal density, which we had derived from electron micrographs (about 4 km/mm^3), as compared to the lower value (1–2 km/mm^3), which we had obtained in the Golgi pictures, since this would imply a much denser spacing of synapses along the axon.

Thus we obtain confirmation of the surprisingly high value for the length of axon per neuron, about 35 to 40 mm. With a spacing of synapses at about 5 μm, the number of synapses per neuron turns out to be between 7000 and 8000 – close enough to the value calculated by the direct comparison of the two densities.

Chapter 10

Statistics of the Synapses Along the Axons

Besides overall density, the distribution of the synapses over the axonal tree can also be studied in Golgi preparations by observing the localization of the axonal swellings or boutons. This was an interesting issue for us, because on the one hand we favour the view of a diffuse network of synaptic contacts between axons and dendrites, implying a homogeneous distribution of the synapses over the whole axonal tree, while on the other hand the results summarized in Fig. 21 seemed to indicate a more preferential localization of the synapses.

The study (Hellwig 1992) prompted by this question referred to 11 primary collaterals of pyramidal cell axons, belonging to three different neurons, and varying in length between 152 and 442 μm. The collaterals with their swellings are shown as idealized straight lines in Fig. 23. The point of attachment of the collaterals to the main axon is to the left on the diagram. While the observation made by Schüz and Münster (1985) on the collateral shown in Fig. 21 seemed to imply a rule which keeps the initial segment of the collaterals free of synapses, this was not apparent in the present sample, even if the collaterals numbered 2, 5 and 11 might have suggested such a rule if taken by themselves. In the entire sample the tendency of the synapses to be denser on the peripheral parts of the axonal tree is no longer apparent; further, the local crowding of synapses which one might envisage in the collaterals numbered 6, 8 and 11 appears rather as a statistical fluctuation. As a matter of fact, very little correlation can be found comparing successive intervals between boutons, if their spacing is subjected to the kind of statistical analysis which has been developed for spike trains (or, in general, "point processes": for references see Aertsen and Gerstein, 1985). The clouds of points in Fig. 24 would have to crowd about the diagonal, if there were regions along the axons with denser or looser spacing of the synapses. This is not the case, the points being well scattered in the plane. The other very sensitive instrument to reveal order in sequences of points along a line is auto-correlation. Clearly, the two correlograms (Fig. 24) obtained from the collaterals numbered 5 and 8 in Fig. 23 reveal hardly any correlation except the trivial and insignificant correlation peak at the point 0. For a full account of this study see Hellwig et al. (1994).

Such apparently purely random spacing of synapses along the axon, "as if it had rained synapses on the axon", may be interpreted in different ways. Strictly speaking, it does not exclude the idea of a very precise

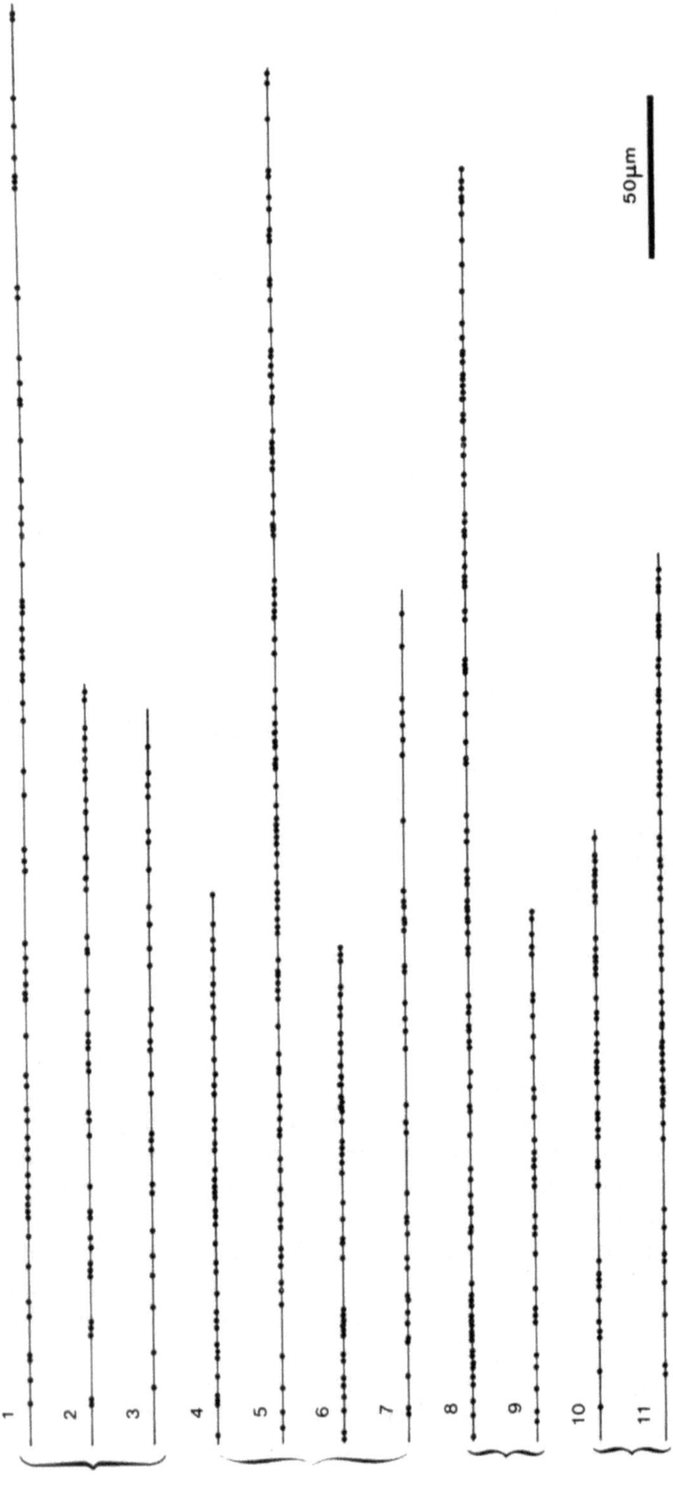

Fig. 23. Primary axon collaterals of pyramidal cells (numbered 1 to 11), straightened to show the distribution of boutons *(dots)*. *Brackets* indicate collaterals which belong to one and the same cell. (Hellwig, 1992)

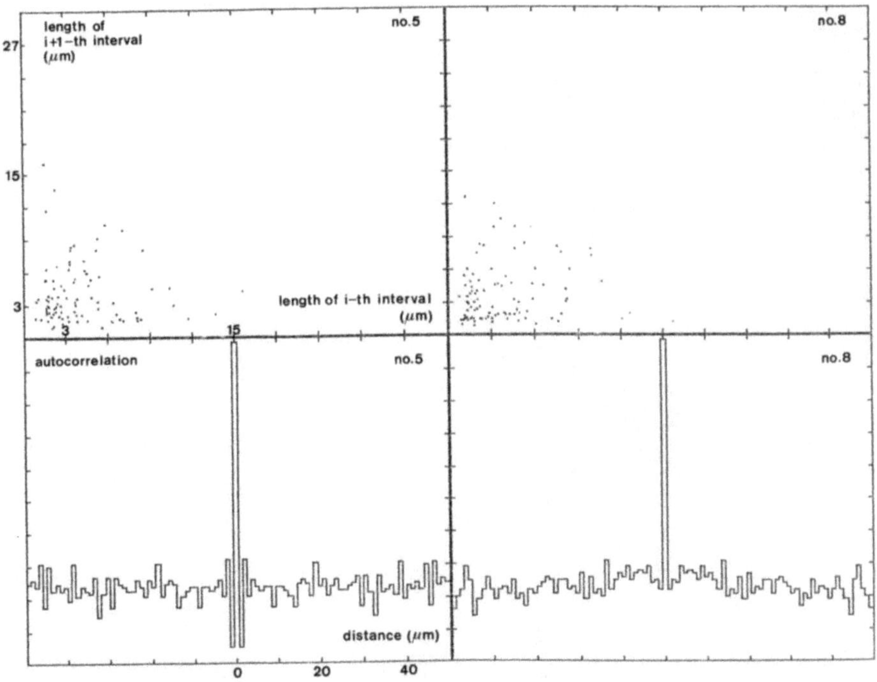

Fig. 24. *Upper row* scatter diagrams showing the length of each inter-bouton interval as a function of the preceding one, for the two collaterals number 5 and 8 of Fig. 23. There is no apparent correlation. *Lower row* the diagram shows the frequency of boutons at various distances given a bouton at position 0, averaged for all the boutons of one collateral (the so-called autocorrelation function). The peak at the position 0 is trivial, the rest is random fluctuation. The symmetry of the distribution has no significance. (Hellwig, 1992)

and detailed "wiring scheme" which might just be too complicated for us to understand or even to see. After all, the structure of a very complicated piece of electronic machinery, if nothing is known about the underlying plan, may look like random wiring and may not reveal any regularity even if sophisticated statistical methods are applied to it. On the other hand, the concept of a "terminal ramification" of the axon, so dear to some neuroanatomists, is not compatible with a diffuse distribution of synapses over most of the axonal tree, such as is the case with the cortical pyramidal cells. So, either the "wiring" in the cortex is complicated beyond all expectation, or it is really quite diffuse, and essentially random, as we prefer to see it.

There is another point. The absence of any periodicity in the spacing of the synapses which the autocorrelograms seem to imply suggests that it is no intrinsic factor that decides when and where an axon makes synaptic contact with the surrounding neuronal elements. More likely it is the complicated dense network of the surrounding dendrites which offers the postsynaptic sites to the axon and decides on their location. It

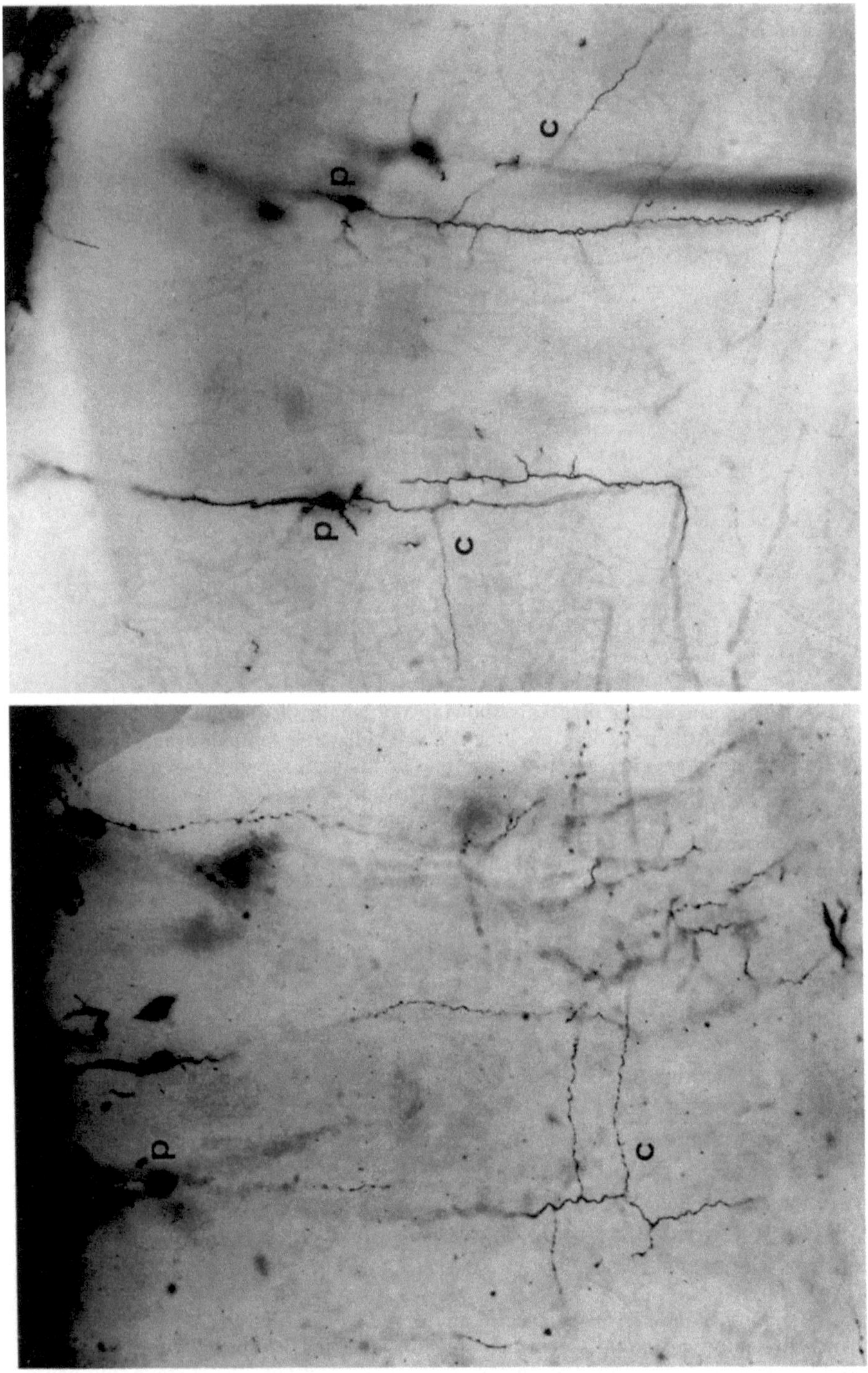

Fig. 25

is known from developmental studies (e.g. Cajal, 1911) that the axons of the larger pyramidal cells in the cortex are developed to a considerable extent when the basal dendrites begin to grow and find their way between them (Fig. 25). The resulting network of intertwined dendrites is to all intents and purposes "random". Thus we may not be too far from the truth when we say that the distribution of synapses along the axons looks *as if it had rained synapses on the axons.*

Fig. 25. Immature pyramidal cells *(p)*. *Left* newborn mouse; *right* 3rd postnatal day. The collaterals *(c)* are well developed, while the basal dendrites have scarcely begun to grow. The apical dendrite and the main stem of the axon are the first to develop

Chapter 11

Density of Dendrites

Quite analogous to the density of axons, the density of dendrites in the cortex can be expressed as length per volume, the total length of all the dendrites contained in 1 mm^3 of cortex. Here again we can start from two kinds of data, electron microscopy and Golgi preparations, and compare the results. The third method which we used in the case of axons, reduced silver staining, is not practicable for dendrites.

We have already seen (Chap. 7) that the dendrites (not including dendritic spines) occupy about the same volume as the axons in the cortex, 29% of the total (35% in the neuropil not including cell bodies and blood vessels, which in turn makes up 84% of the entire volume). The dendritic length can be obtained by dividing the volume occupied by the dendrites through their cross-sectional area. Assuming a cylindrical shape and an average diameter of 0.9 µm (measured on electron micrographs), the dendritic length in 1 mm^3 turns out to be 456 m. This is almost a whole order of magnitude less than the axonal length contained in the same volume and reflects the greater average thickness of the dendrites.

The dendritic length can also be measured directly in Golgi preparations. With the value of 3 to 5 mm of dendrites per neuron which we had assumed in our earlier papers (Braitenberg, 1978b), the 9.2×10^4 neurons in 1 mm^3 would correspond to between 276 and 460 m of dendrites in 1 mm^3. This is in good accord with the figure obtained from the electron micrographs.

Similarly, the number of synapses on the dendrites fits our estimates of dendritic length. If the length of the dendrites in the tissue is one tenth of the length of the axons, and if one postsynaptic element corresponds to each presynaptic one, then we expect ten times more synapses per unit length on the dendrites than on the axon. We have already calculated (and confirmed by direct observation), the value of one synapse every 5 µm for the axon. Thus we expect two synapses per µm on the dendrites. We shall see later (Chaps. 22, 23) that this is very close to the measured value.

Chapter 12

Different Kinds of Synapses

Various biochemical assays demonstrate a variety of neurons and of synapses suggesting different roles in their functional interplay. These different roles may well coincide with the two fundamental interactions postulated by neural net models, excitation and inhibition, and with the other distinction which such models sometimes make between fixed and modifiable synapses. The functional interpretation of the biochemical variety remains hypothetical for the most part, although the identification of a certain transmitter (glutamate) with excitatory interactions and of another (GABA) with inhibition seems to be generally accepted. We have not done any cytochemistry ourselves but follow the standard usage of the distinction between two types of synapses according to their appearance in the electron micrographs. They are widely thought to correspond to excitatory and inhibitory interactions respectively. The distinction of two types of synapses was introduced by Gray (1959) and later refined by Colonnier (1968). Uchizono (1965) was mainly responsible for their interpretation in terms of excitation and inhibition on the basis of the well worked-out anatomy and physiology of the cerebellum. The two kinds of synapses, which we will call Type I and Type II following Gray (1959) are thus characterized (Fig. 26):

Type I: The pre- and postsynaptic membranes are more heavily stained than the non-synaptic membranes, and in addition "electron-dense" material, i.e. some substance stainable with Osmium, is attached to the postsynaptic membrane. Inside the presynaptic cell of such synapses, in the immediate vicinity of the specialized membrane, there are many vesicles of a fairly uniform round shape.

Type II: As in the Type I synapses, the synaptic membranes appear darker than the surrounding membrane. However, there is no electron-dense material attached to the postsynaptic membrane inside the postsynaptic neuron (Colonnier, 1968). The vesicles on the presynaptic side tend to be of an irregular elongated shape. The synaptic cleft is usually narrower than in the Type I synapses.

The emphasis which different authors put on the different criteria for the distinction of the two kinds of synapses varies, and it has even been claimed that both types are nothing but caricatures of the extremes in a continuum, calling into question the entire business of a distinction of discrete types of synapses. We believe that Type I and Type II synapses are a reality, and that if there are transitional forms between them, they

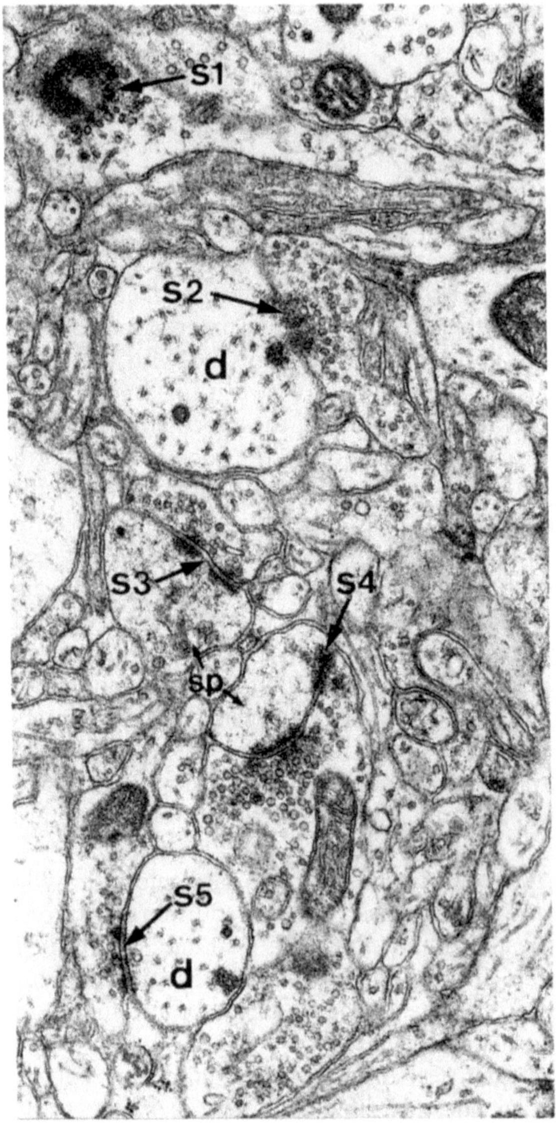

Fig. 26.
Synapses in the mouse cortex, electron micrograph. *S1* synapse cut tangentially; *S2* synapse between an axon and a dendritic shaft, slightly obliquely cut so that one cannot decide whether it is of Type I or Type II; *S3* synapse of Type I, with a spine head as the postsynaptic element; *S4* Type I, on spine head; *S5* synapse of Type II on a small dendrite. *d* dendrites; *sp* spines. Enlargement × 23 500

are only apparently so, perhaps due to unfavourable orientation of the synapse in the section obscuring the real geometry, or to insufficient preservation of the tissue. The criteria which we have mentioned are those which we used in our counts. However, when we were in doubt, we gave priority to the criterion of the existence or non-existence of a postsynaptic thickening. This seemed to provide the basis for a secure diagnosis in 98% of the cases.

For our differential counts of Type I and Type II synapses only those synapses were considered whose synaptic junctions were oriented perpendicularly to the plane of the section, the indication of this being the synaptic cleft clearly visible between the pre- and postsynaptic membranes. On such specimens the distinction of the two types is greatly facilitated.

The percentage of Type II synapses is shown in the following table for the three mice and the three cortical areas in which neurons and synapses were counted.

	Area		
	8	6	17
Mouse 1	7 (12)	11 (14)	9 (13)
2	4 (5)	6 (7)	8 (8)
3	8 (10)	12 (13)	9 (10)

The figures in brackets are obtained by including among the Type II synapses all the cases in which the diagnosis was uncertain.

The percentage of Type II synapses can thus be assumed to be between 8 and 10%, including the debatable cases. The figures for mouse 2 are puzzling, since they are consistently lower than the others. We may assume a fault in the preparation of the specimens, but this would lead us to the strange conclusion that bad fixation or staining of the tissue mimicks the characteristics which we have used to identify Type I synapses, rather than those of Type II. It may of course be that there are individual differences in the relative frequency of the two types of synapses, perhaps reflecting different frequencies of different kinds of neurons. We shall resist the temptation of relating inter-individual differences in the proportion of excitatory and inhibitory synapses to the fact that some brains produce epileptic fits more readily than others (mouse 2 was not noticed to be in any way pathological in its behaviour).

These counts were again made on regions of the cortex representing "pure neuropil", i.e. without cell bodies and without blood vessels. The correction which we have to apply for this in the case of the synapses might be expected to change the statistics, since it is known (as we shall discuss later) that Type II synapses are localized preferentially on the cell body. We therefore expect Type II synapses to be underrepresented in our sample. But the effect is surely not very noticeable since we know from various studies (Müller et al., 1984, on the rabbit; White and Rock, 1980, on the mouse; Peters and Kaiserman-Abramof, 1970, on the rat) as well as from our own counts that the number of synapses on the cell body rarely exceeds 200 and is usually reported as being much smaller. Granted the higher number and supposing they are all type II, the synapses on the cell body amount to only 2.5% of the total, causing the percentage of the Type II synapses to rise to about 11%.

The results obtained by other authors (Peters and Feldman, 1976; Wolff, 1976; Bär, 1977; Blue and Parnavelas, 1983 on the rat; Winfield,

1983; Beaulieu and Colonnier, 1985, on the cat) vary between 5% and 21% Type II synapses.

On the same electron micrographs as used for our differential counts of Type I and Type II synapses we could also establish the frequency of another characteristic of synapses; namely their localization on the heads of dendritic spines rather than directly on the shafts of the dendrites. This is not simply a third type of synapse at the level of the classification in Type I and Type II since the synapses on the heads of the spines are almost all of Type I (see Chap. 22). It is rather a differentiation within the class of Type I synapses, an important distinction which we shall tentatively relate to the difference between modifiable and non-modifiable synapses in a later chapter (Chap. 24). The head of a spine (not the neck) can be easily recognized on the electron micrographs (Figs. 26 and 42) when a sufficient portion of it is contained in the section. This is always the case when we have a synapse large enough to be included in our counts and well enough orientated in the section to be recognized as Type I. As has been pointed out by many authors (e.g. Peters et al., 1970), the cytoplasm within the spine head contains some diffuse material stained with Osmium, which gives it a peculiar appearance, and it does not contain mitochondria or any tubular structures. For further detail see Chapter 22 and Fig. 42.

According to these criteria, 77% of the synapses in the neuropil resided on heads of spines. It will be remembered that the regions around the cell body were not included in these counts and that the figure must therefore be corrected for the 2% of synapses localized there. Thus about 3/4 of all synapses in the cortex have spine heads as their postsynaptic element. With 75% synapses on spines (nearly all of them of Type I) and 89% of all synapses in the neuropil being Type I, we deduce that (nearly) 85% of Type I synapses are localized on spines.

Dendritic spines are obviously important components of the cortical machinery and we are not surprised to find that 14% of the volume of the neuropil is occupied by spines (heads plus necks).

Chapter 13

Interim Discussion

We have now gone as far as we can go with our considerations of the cortex as a purely statistical ensemble of axons, dendrites and synapses. It was our contention that the comparison of such quantities as axonal and dendritic density, density of spines and density of synapses would yield some ratios which could be interpreted in terms of cortical function, even before there was any mention of the shape of neurons and of their specific connections. Actually, we believe that some theoretical models of the cortical wiring are already excluded by the data at this stage.

A few examples. Eight thousand synapses on each neuron are an impressive number, but even in the small brain of the mouse it is obvious that a scheme which requires direct connections between all the neurons of the cortex, as some models of an "associative matrix" kind would have it, is out of the question. The number of synapses on each neuron would have to be more than 1000 times greater to provide individual connections to the 10^7 neurons of the mouse cortex.

There are models which assume an equal number of positive and negative interactions between the neurons, or in terms of physiology, the same number of excitatory and inhibitory synapses. This may be a convenient assumption if you think of a logical network embodying conjunction and dysjunction of negated and not negated terms. It may also be an obvious assumption if you think of economic coding in the framework of some information theoretical view of the cortex. A symmetric distribution of positive and negative influences between the neurons may also be an obvious assumption in some mathematical contexts in which physicists feel at home, such as wave propagation and interference. But it is obviously not what the histology reveals, if we take the 89% Type I synapses as mediating positive influences and the few Type II synapses as mediating the negative ones, which is to say, Type I = excitatory and Type II = inhibitory, as we all believe.

Eight thousand synapses on each neuron are too many if we are to see neurons as "threshold devices", as units which respond "all or none" to specific constellations of activity in their input lines. If the threshold is to be set exactly at some number between 0 and 8000 in order to define the set (or sets) of active input lines to which the neuron is supposed to respond, we have to assume an immaculate precision of the electrochemical phenomena in the neuron membrane which no physiologist would deem possible.

There is yet another view of the cortical "wiring" which can hardly be upheld in the face of the facts we have collected up to now. This is the idea of precisely addressed connections from individual neurons to other individual neurons. The diffuse and apparently haphazard arrangement of synaptic boutons along the branches of the axonal tree suggests rather a distribution of signals to large sets of other neurons which happen to stretch their dendrites in the direction of that particular axonal tree. The point is not absolutely conclusive, however, and may not convince those among our colleagues who are sold to the idea of "specific connections". Among other things, our measurements refer to the mouse cortex, where there are hardly any myelinated fibres especially in the upper layers. The presence of many such fibres in other species would certainly show in the overall distribution of synapses on the axons. As long as we have no better explanation for the myelin which covers some of the intracortical axons, the increase in "specificity" which comes with the insulation of some axonal segments remains a valid explanation.

However, if there were a majority of axonal segments without any synapses, the density of synapses on the remaining segments would exceed by far the overall average. We have never observed such local crowding of synapses on axons.

Thus, on the whole, we have already shifted the balance away from the idea of "wiring", exemplified by radio or computer engineering, toward that of a network of an entirely different kind, set up largely by chance and possibly refined by learning processes. Mind you, this refers to the cerebral cortex and may not be valid for other parts of the brain! In the brain stem of vertebrates some of the connections are undoubtedly much more specific, and in some invertebrate networks, e.g. in the visual system of the fly, individual fibres are connected with individual target neurons in accordance with an absolutely unerring scheme.

On the other hand, even if much of the synaptology in the cerebral cortex is left to chance, and even if this principle is related to the very essence of the cortical function (as we intend to show later), it would be wrong to think that there is no order in the cortical network. There are statistical constraints to the randomness of connectivity, and these constraints are given by the form of the various kinds of neurons which are found in the cortex. If we talk about neurons as clouds of presynaptic sites, represented by thousands of synapses on the axon, intermingling with clouds of postsynaptic sites, represented by the synapses on the dendritic trees of other neurons, what we have in mind is a statistical view of the functional interactions in the cortex. But the likelihood of a connection, the number of synapses in the neuropil are questions for which answers can only be found in the geometry of the dendritic and axonal trees of different neuron types.

Thus we have to turn next to a classification of cortical neurons, and then consider their individual contribution to the statistics of neuronal connection in the cortex.

Chapter 14

Morphology of Neurons: Golgi Pictures

As soon as the Golgi technique made it possible to stain individual neurons in the tissue, dissecting them out, as it were, from the tangle of their intermingled processes, it became apparent that they come in a spectacular variety of shapes and sizes. The collection of illustrations in Cajal's Histologie (1911) together with further drawings only recently made available in a translation of Cajal's original papers (DeFelipe and Jones, 1988) furnishes a panorama of this variety to which more recent publications (see papers in Peters and Jones, 1984a onward) have added comparatively little. What improved staining techniques have shown recently, however, particularly by means of dyes injected into the cell (Gilbert and Wiesel, 1979, and many others; see Parnavelas, 1984), is that the Golgi stain may occasionally leave part of the axonal tree unstained. We have already noted this when comparing our estimates of the axonal density obtained from Golgi pictures with those inferred by other means (Chap. 7). The general picture, however, has not changed much since Cajal and the old dilemma between an urge to differentiate neurons in an ever finer taxonomy, and the opposite urge to reduce the variety to a few simple categories is still with us today. There seemed to be no way of determining whether the "splitters" were more right than the "lumpers" when we started on the mouse cortex, and this was one of the reasons why we decided to produce our own collection of Golgi preparations. Today there is electron microscopy (Colonnier, 1981; LeVay, 1973; Peters, 1984a, b; Somogyi, 1978), which seems to confirm the intuition of those who preferred a parsimonious classification (as in Globus and Scheibel, 1967; Braitenberg, 1978b), but the puzzle of the variety of shapes of cortical neurons still remains unresolved and will have to await a better understanding of the network.

We have Golgi preparations of over 200 adult and 45 immature mouse brains. In addition we have in our collection 10 Golgi-stained rat brains and 38 guinea pig brains which we used for comparison. The techniques we used for Golgi staining were, in order of preference: Colonnier's (1964) method combining dichromate and aldehydes as fixatives, the time-honoured Rapid Golgi procedure (Osmium and dichromate), so successful in the hands of Cajal, and in a few cases the Golgi Cox method.

We did, of course, experiment with these methods, applying slight variations here and there, but usually reverted to the classic prescriptions. There was only one improvement which we would like to hand on to future users of the Golgi methods. In our search for ways of managing the selectivity of the Golgi stain in such a way as to make it comparable to modern stains by injection, we experimented with local application of the silver nitrate (after the inevitable dichromate fixation) by small injection needles, by micropipettes, by the insertion of $AgNO_3$ crystals and by capillary diffusion through small wicks. It turned out that positioning the dichromate-fixed brain on top of a stack of filter paper wet with $AgNO_3$-solution, rather than immersion in the solution, would always produce distinct staining of some neurons near the site of contact, and very little staining of neurons farther away. The quality of the stain, as well as the clarity of the background, was also improved by this trick. Especially where the Golgi stain was combined with electron microscopy, the trick with the wet filter paper was invaluable.

Chapter 15

Classification of Cortical Neurons

There is no justification beyond recourse to esthetic motives for the preference of one classification over another and all classifications are likely to stumble over exceptional cases. In our earlier papers (Braitenberg, 1977, 1978a,b), far from being discouraged by the ever multiplying zoo of cortical neurons described in the literature, we proposed a distinction between two main classes, pyramidal and stellate cells[1], with a third class, Martinotti cells perhaps occupying an intermediate position, but well enough characterized to be kept separate. The classes were defined as follows.

Pyramidal Cells. The dendrites, beginning a short distance from the cell body, are covered with many spines, at least one spine every micrometer of dendritic length. The main stem of the axon almost always takes a vertical downward course, with little deviation from a straight line. The same tendency can be seen in the primary and secondary branches of the axon, which may take off at any angle but tend to keep their straight course from one point of ramification to the next, or to the point of termination. The branching pattern of the axon is very loose: the entire ramification occupies a large volume. The main axon enters the white matter and either projects to another region of the cortex or to a different part of the brain, or to both.

Most pyramidal cells have a clearly bipartite dendritic tree, with basal dendrites distributed stellate fashion around the cell body, and an apical dendrite pointing upward and usually ramified in the upper layers of the cortex. Some pyramidal cells have hardly any apical dendrite, or have the apical dendrite confused with the basal dendrites, as may happen when the cell body is located in the upper tiers of the cortex. This has led some authors to the definition of "spiny stellate cells". It was our impression that the continuous transition of dendritic form between true pyramidal cells and spiny stellate cells justifies the inclusion of the latter in the general class of the former.

[1] The term stellate cell has been used ambiguously in the past and is therefore avoided by many authors. For lack of a better term we shall use it here, fully aware of the fact that some of the cells we include in this category do not have the star-shaped dendritic tree which is implied by the Latin term.

The defining characteristics of the pyramidal cells are then: spines on the dendrites, loose axonal ramifications with a rather straight course and a main axon leaving the grey matter. Thus, the pyramidal cells make both, short-range and long-range connections.

Stellate Cells. The dendrites are without spines. If there are any excrescences on the dendrites resembling the spines on pyramidal cells, their number is small, less than one every 5 μm of dendritic length. The arrangement of the dendrites around the cell body is usually stellate, i.e. without any preferential orientation, the dendrites being equally likely to point in any direction. In some varieties there is a bunching of the dendrites around the vertical direction. We do not make a special type out of this variety, as some authors do.

The pattern of the axonal ramification of stellate cells varies a great deal and may well justify a distinction of several classes within the general type (see Fig. 85, Chap. 36). But it is so different in every case from that of the pyramidal cells that one cannot be confused with the other even if only a few branches are seen in the microscope. The axon has no well-defined main stem and does not leave the gray matter, as in the case of pyramidal cells. The axon may leave the cell body in any direction. It ramifies over and over again, the distances between successive branching points being often 10 μm or less. The ramification is very dense. It occupies a region no more than a few hundred micrometer across[2] which often overlaps generously with the region in which the dendrites of the same neuron are distributed.

Thus smooth dendrites and a dense axonal tree which makes only short-range connections are the prominent features of stellate cells in the Golgi picture.

Martinotti Cells. The dendrites of the third type of cortical neurons carry spine-like excrescences (it has never been shown in the electron microscope whether they are really related to the true spines) intermediate in number between those of pyramidal cells and those of stellate cells. The pattern of the distribution of dendrites around the cell body varies, being sometimes confined to a sphere, sometimes elongated in a vertical or in a horizontal direction. The axon leaves the cell body in a vertical upward direction, opposite to that of the pyramidal cells. The axon has a well-defined main stem from which collaterals depart; these may have secondary, tertiary and even higher-order branches. The collaterals take a strangely scraggly course, very different from the straight branches of the pyramidal axon and different also from the more smoothly winding course of the stellate axon branches. Our tendency to define a third main type of cortical cell, the Martinotti cell (thus named after an early refer-

[2] With exceptions, see Fairén et al. (1984).

ence, Martinotti, 1889) is based on the repeated experience that following back a particularly scraggly branch of axon in the preparation, we were led to its origin from a straight ascending axon.

Thus to our mind the ascending main axon, the irregularly winding (scraggly) branches of the axon and, to a lesser extent, the few spine-like appendages on the dendrites define the Martinotti cell. There are perhaps two types of Martinotti cells distinguished by their location, upper (or short axon) Martinotti cells in the middle layers, and lower (long axon) Martinotti cells in the bottom layers of the cortex.

The drawings in Figs. 27, 28, 29 show some examples of each of the three types of cortical neurons. Care has been taken not to select the typical textbook cases, but to give an impression of the range of variation that is normally encountered.

Figure 30a, b, c shows the localization of pyramidal cells, stellate and Martinotti cells in a survey of our Golgi preparations (Staiger, 1984). The cell types were classified according to our definitions. Not in all cases were all the criteria available, since neurons with incompletely stained axons, or even with incompletely stained dendrites were also recorded as long as they provided satisfactory evidence for their inclusion in one or the other category (richly spiny dendrites for pyramidal cells, dense axonal ramification for stellate cells, straight upward axon for Martinotti cells etc.). Only neurons located within the dorsal half of the hemisphere, i.e. roughly above the rhinal sulcus, were recorded, and their position was marked on the outline of the dorsal view of the hemisphere from the SAP atlas. For the determination of the coordinates of each individual neuron, the section in which it was found was matched as closely as possible with the corresponding section of the atlas, and then the coordinates were transferred to the maps.

The three neuron types reveal no clear pattern in their distribution over the hemisphere, except perhaps for some preferential zones in the case of the Martinotti cells. Their number, however, was not sufficiently large to exclude accidental effects.

The relative frequency of the three types is an open question. In Nissl preparations, where every cell is stained, it is almost impossible to distinguish the cell bodies of pyramidal cells from those of the other types (claims to the contrary by Werner et al., 1979). The Golgi preparations, on the other hand, may be very biased in their selectivity for different cell types. Thus, for example, in our own preparations Martinotti cells are quite frequently present, while they seem to be entirely missing in the collections of some of our colleagues who also use the Golgi method. On the other hand, we have never seen, among several thousand neurons stained, a convincing specimen (at least not a complete one) of the "chandelier" kind, which is a well-documented variety of stellate cell (Szentágothai, 1975; Tömböl, 1978; Fairén and Valverde, 1980; Peters et

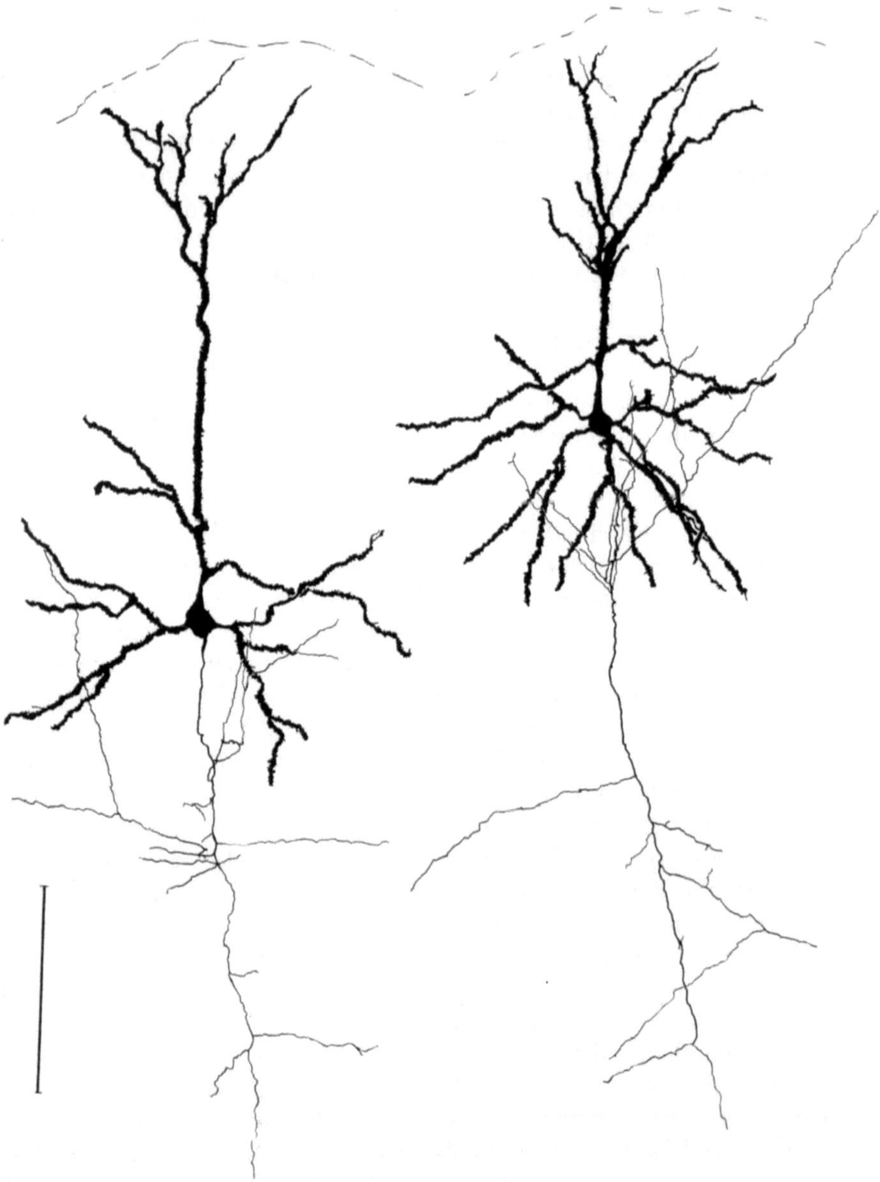

Fig. 27. Camera lucida tracings of three pyramidal cells, Golgi stain. Bar 0.1 mm. Dashed: surface of cortex. (Braitenberg, 1978b)

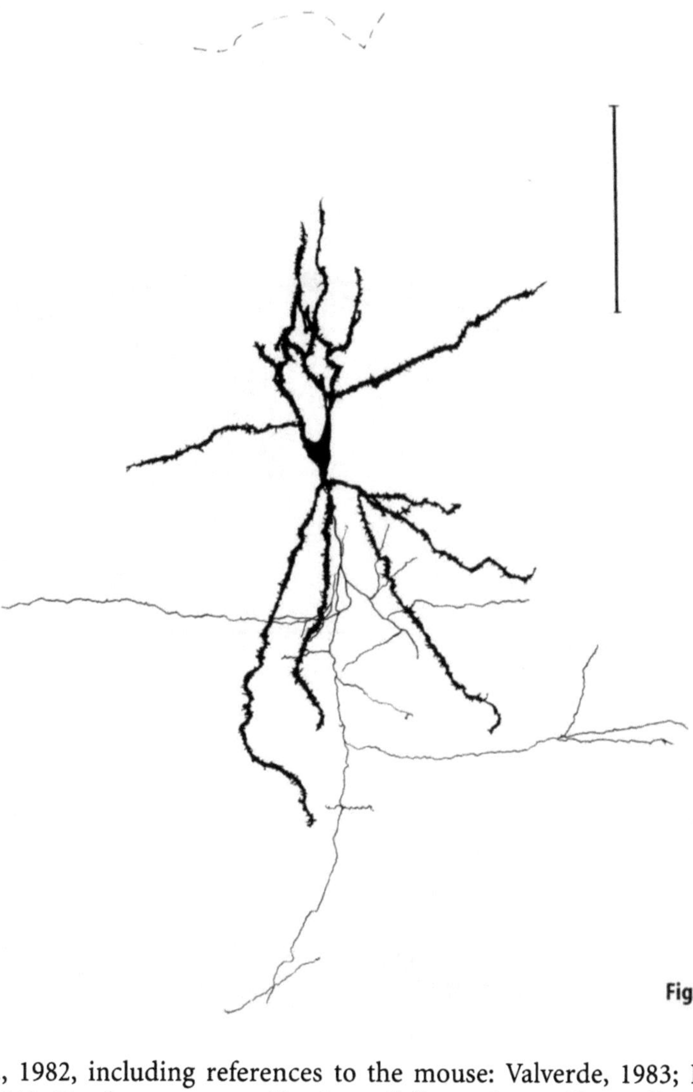

Fig. 27 (continued)

al., 1982, including references to the mouse: Valverde, 1983; Fairén et al., 1981). Nor have we seen "basket cells", another stellate cell variety, which is, however, less clearly set off from the rest. The "baskets" which their axons are supposed to form around the cell bodies of other neurons (first illustrated by Cajal, 1911) are now reputed to be composite, each being formed by the collaterals of several cells of the ordinary stellate kind coming together on one pyramidal cell body (Marin-Padilla, 1969, 1972). Here apparently the tufts of terminal branches with which each collateral claws the cell body of one Purkinje cell in the case of the basket cells of the cerebellum, have inspired one of the few false perceptions by Cajal and consequent misnomers by others.

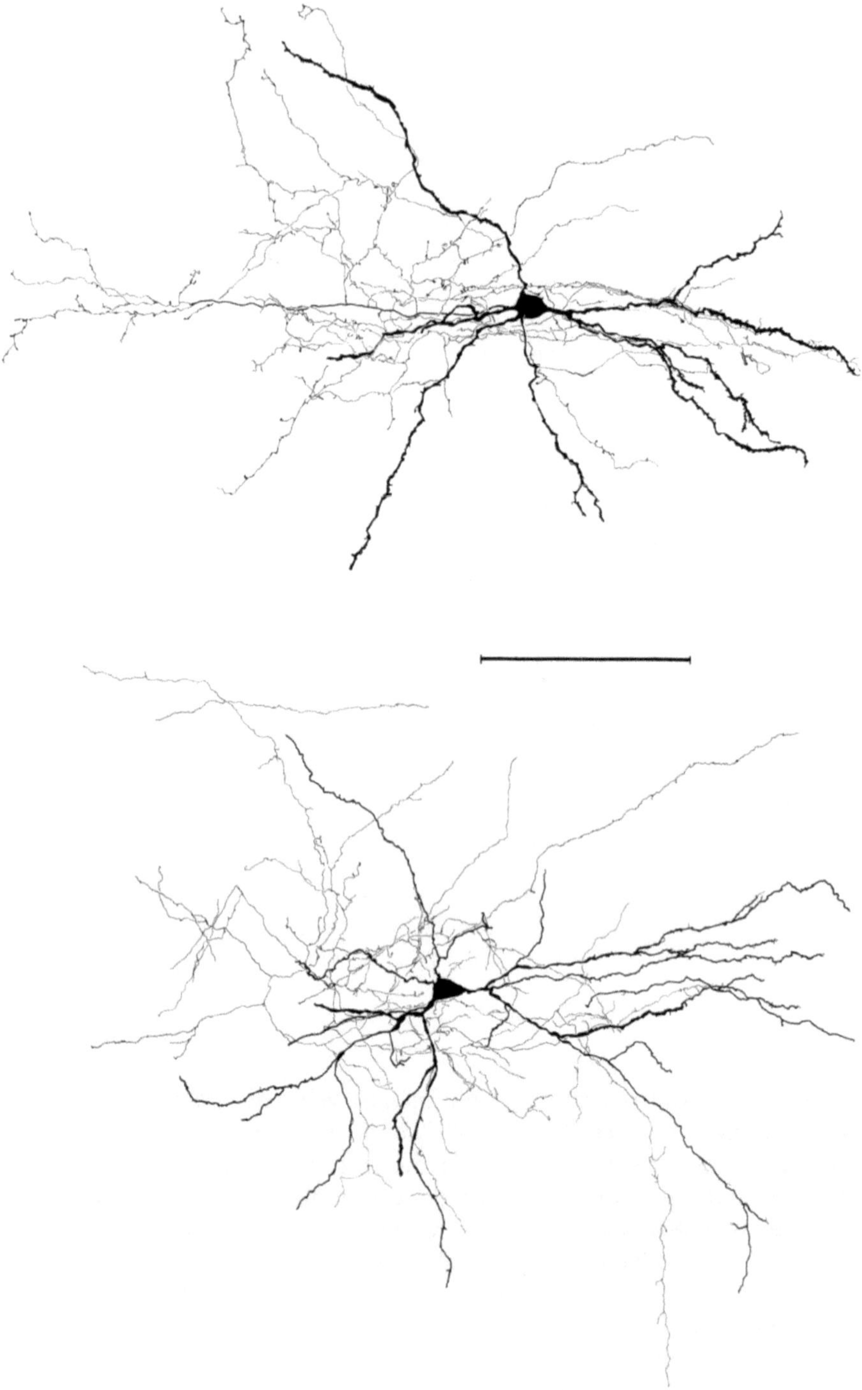

Fig. 28

Estimates of the relative frequency of pyramidal and non-pyramidal (= stellate and Martinotti in our terminology) cells may be based on Golgi-Cox preparations which are supposed to provide a more representative selection of the different types than the other Golgi techniques (Mitra, 1955), or may be underpinned indirectly by differential counts of synapses in the electron microscope (Winfield et al., 1980; Peters and Kara, 1985). The reasoning behind the latter procedure will become clear once we have described the distribution of synapses on the neurons of the different types. The estimates for mice and rats vary between 62 and 85% pyramidal cells. Our own estimate lies close to the higher value and counts on human lipofuscin stained material also yield 85% pyramidal cells (Braak and Braak, 1986). We have no way of estimating the relative frequency of the other two neuron types. However, judging from the small distance at which we see occasionally two neighbouring Martinotti cells in our preparations, we believe they are not as infrequent as the little attention which they receive in the literature would imply.

We are aware of the risk inherent in our classification, which was primarily based on purely morphological criteria. The obvious pitfall is oversimplification due to neglect of a variety which may be only apparent at the biochemical level. However, since we first adopted our scheme a great deal of evidence has come to light, mainly based on combined Golgi staining and electron microscopy, which confirms at least one of our intuitions, namely the special status of the neuron with densely spiny dendrites, the pyramidal cells in our terminology, and the inclusion in that class of at least some of the so-called spiny stellate cells (Lund, 1973).

The question of classification of cortical neurons from a modern point of view has been discussed by Peters and Jones (1984b) in a review which introduces a thorough treatment of the various classes in a series of articles by Feldman (1984), Fairén et al. (1984), Lund (1984), Jones and Hendry (1984), Peters and Saint Marie (1984), Marin-Padilla (1984) and Tömböl (1984). The well-known volume in which these articles are collected, together with its companion volumes and White (1989), make it unnecessary for us to go into details. While more observations are being made with ever new techniques on different neuron types in a variety of animal species, the most important statements seem to hold true. They are, according to Peters and Jones (1984b):

Pyramidal cells (including those with no apparent apical dendrite, sometimes classified as spiny stellate) can be recognized in electron micrographs because their cell body is postsynaptic to synapses of Type II

◄

Fig. 28. Camera lucida tracings of two stellate cells of the mouse cortex. Golgi stain. *Thin fibres* axons, *thicker fibres* dendrites. Note the dense axonal ramification occupying nearly the same region as the dendrites. Bar 0.1 mm. (Braitenberg, 1978b)

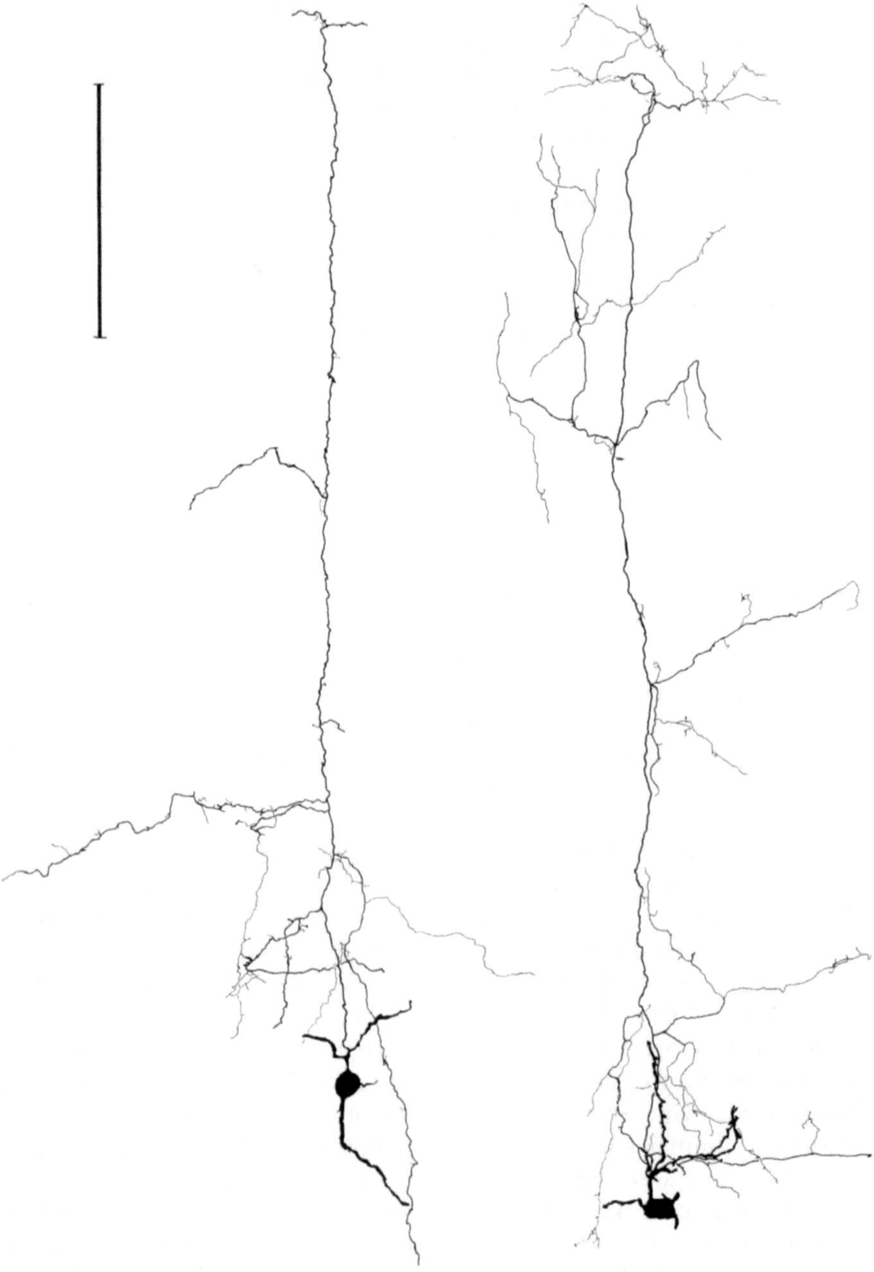

Fig. 29. Camera lucida tracings of four Martinotti cells of the mouse cortex. The dendrites *(heavier lines)* are remarkably short. The main axon is ascending and reaches the uppermost layer of the cortex in three of the four neurons. Bar 0.1 mm. (Braitenberg, 1978b)

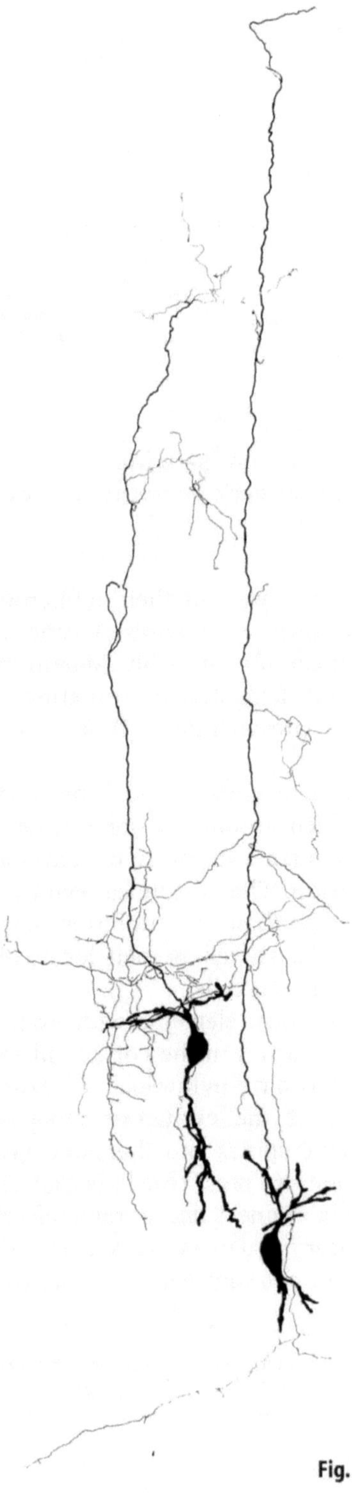

Fig. 29 (continued)

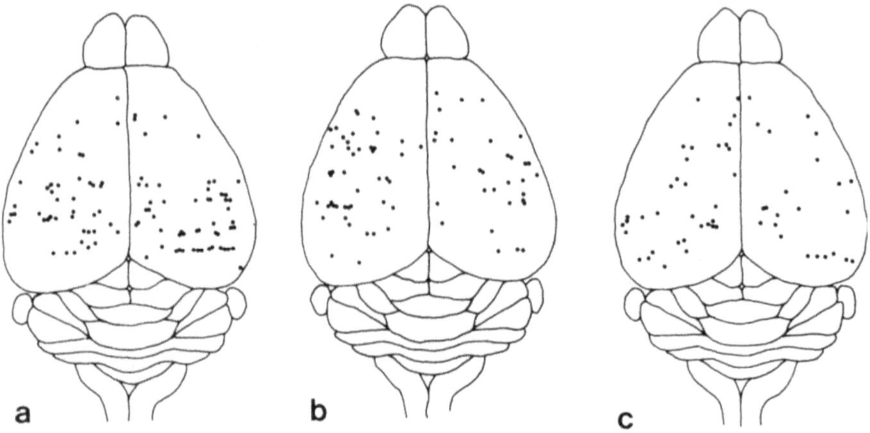

Fig. 30 a–c. Localization of pyramidal cells (a), stellate cells (b) and Martinotti cells (c) in a sample collected by V. Staiger from our Golgi preparations. No preferential localization is evident on the dorsal surface of the hemispheres

exclusively. Pyramidal cells receive Type I synapses on their numerous spines. The axon of pyramidal cells is presynaptic to synapses of Type I.

On the contrary, non-pyramidal cells, including probably Martinotti cells, are postsynaptic, on their cell body and dendrites, to a mixture of Type I and Type II synapses. Their axons are presynaptic to Type II synapses.[3]

In other words, assuming that Type I means excitatory and Type II inhibitory synapses, a conjecture which has been around for some time is confirmed: the cortex is composed of two main categories of neurons, one exerting excitation and the other inhibition. The anatomical evidence strongly suggests that excitation travels farther than the inhibition since on the average the axons of the pyramidal cells as well as their local collaterals are longer than those of the non-pyramidal cells.

The pyramidal cells being the only intracortical elements with excitatory axons[4], it follows that most excitatory synapses in the cortex (all except the ones served by extracortical afferents) have pyramidal cell axons as their presynaptic elements. Since most of the excitatory synapses (85%, see Chap. 12) are localized on dendritic spines and therefore have again pyramidal cells on the postsynaptic side, we must conclude that the majority of excitatory synapses in the cortex connect one pyramidal cell to another pyramidal cell. And since excitatory synapses are almost 90% of all cortical synapses, it turns out that *the dominant feature of the cor-*

[3] With respect to Martinotti cells, there is now enough evidence showing that they are also inhibitory (Hedlich et al., 1990; Wahle, 1993; Kawaguchi and Kubota, 1996)

[4] Disregarding, for the time being, sporadic evidence to the contrary: Peters and Kimerer (1981)

tex is the pyramidal-cell-to-pyramidal-cell connection, represented by 75% of all synapses.

This is an important proposition that will detain us for the rest of this book. It is the definite blow to the idea of a "circuit" as a device in which elements of different kinds are connected to fulfill some shrewd computing scheme. The synaptic relations within the cortex are mostly between elements of one kind.

Chapter 16

Quantitative Aspects of the Three Types of Neurons. Methods

The reason why we first measured the length of axonal branches of individual neurons in the cortex, was our impression that the density of the axonal tree is very different in different neurons and may be used as one of the distinguishing characteristics of pyramidal and non-pyramidal cells. This was indeed confirmed, the pyramidal cells producing the least dense axonal ramification and some stellate cells, as well as some extracortical afferents (Fig. 31) the densest in the cortex.

Our measurements of axons and dendrites were made on Golgi-stained neurons projected with a camera lucida at a magnification of 950 diameters. Since the thickness of the sections was mostly between 60 and 100 μm, the axonal ramification of a neuron, which often measures several hundred micrometers, and even the dendritic tree was only in part contained in the section. For the measurement of axonal (and dendritic) length we had to apply a correction to compensate for the part of the axonal (dendritic) tree that was cut off, and another correction was necessary in order to compensate for the foreshortening of the fibres in the camera lucida projection, depending on the angle they form with the image plane.

I. The average shortening, if all angles are equally likely, is the average sine of the angle α between the fibre and the line of sight, taken from the parallel to the perpendicular orientation.

$$\frac{1}{\pi/2}\int_0^{\pi/2} \sin\alpha \, d\alpha = \frac{2}{\pi} \,. \tag{9}$$

Thus the average length of the segments measured on the projection has to be multiplied by $\pi/2$ to obtain their true average length.

IIa. The other correction was made on the assumption that the ramification of a neuron, including the part cut off, in reality is distributed in a sphere with a radius equal to the longest branch of that neuron contained in the section. The total length of the ramification is then to the length measured in the section approximately as the volume of the sphere is to the volume of a cylinder with the same radius and height equal to the thickness of the section (Fig. 32a). This is $4/3 \times r/d$, where r is the radius of the sphere (and of the cylinder) and d is the thickness of the section.

IIb. Sometimes, when the ramification in question has axial rather than spherical symmetry, with the axis parallel to the plane of the sec-

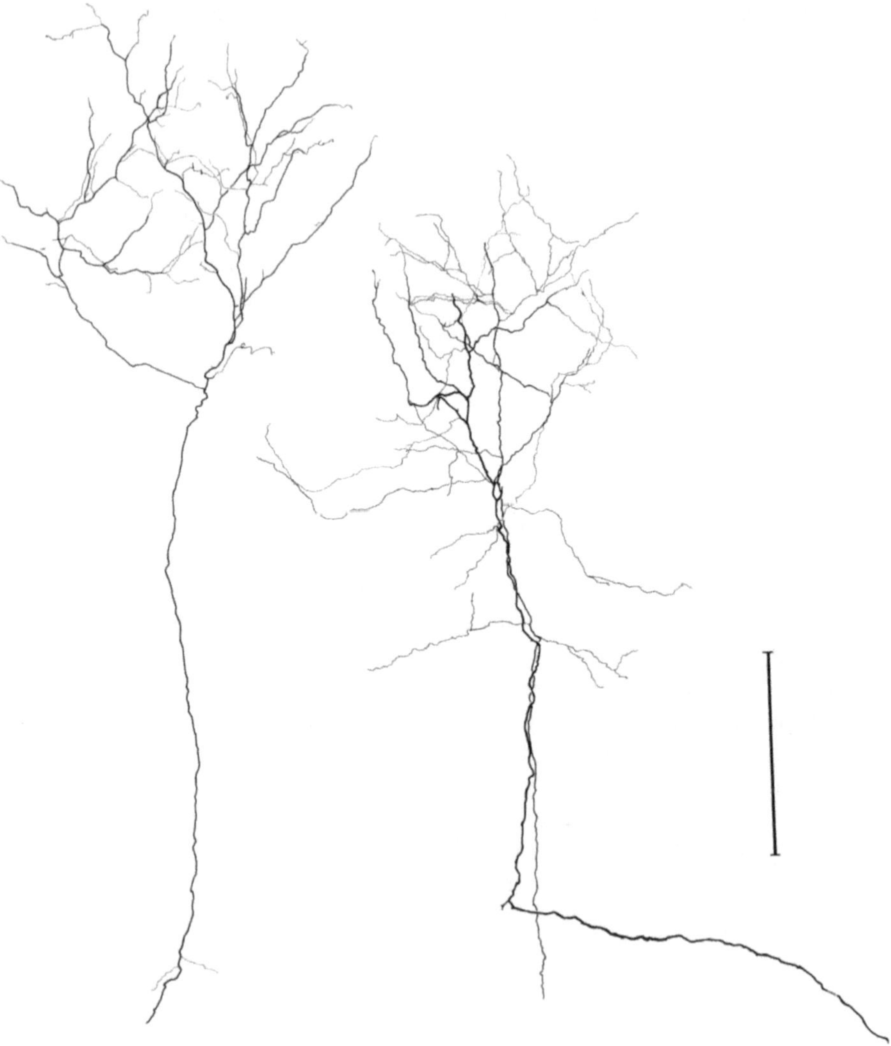

Fig. 31. Thalamocortical afferents with terminal ramifications in the middle layer of the mouse cortex. Golgi stain. On the *right* there are two fibres running very close to each other for much of their course. Their terminal ramifications are nearly coincident. Bar 0.1 mm. (Braitenberg, 1978b)

tion, a more fitting correction can be made by reference to Fig. 32b. The factor turns out to be (volume of the cylinder with radius *r*/volume of a box of the same length with sides $2r$ and d) $\pi/2 \times r/d$.

These were the two corrections which were applied to the measured lengths of fibres on camera lucida drawings in order to obtain the figures reported in Braitenberg (1978b). In the following years a larger number of neurons was drawn in our laboratory and the length of axons and den-

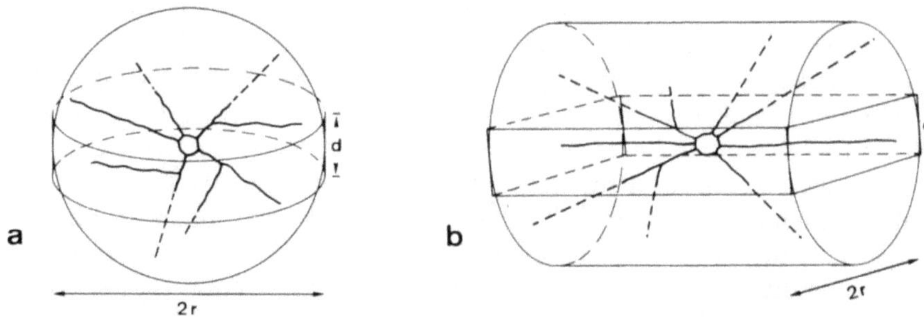

Fig. 32 a, b. To illustrate the way we inferred the size of the dendritic or axonal tree of a neuron from the part measured in a histological section of thickness *d*. Spherical symmetry of the ramification is assumed in **a** and cylindrical symmetry in **b**. *r* is the radius of the sphere or cylinder, equal to the longest fibre measured in the preparation

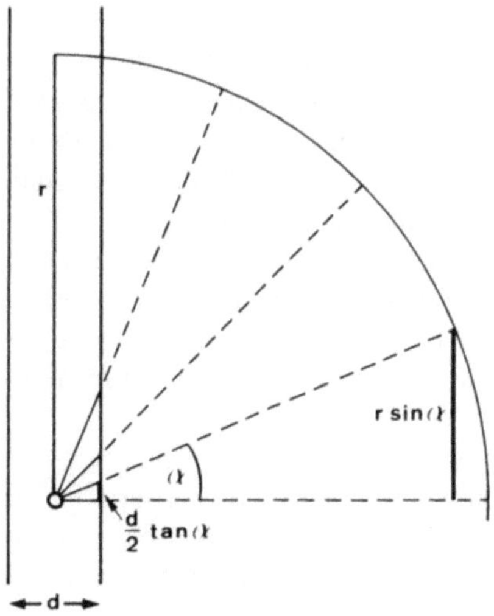

Fig. 33. Fibres of equal length *r* emanating from a neuron located in the middle of a slice of thickness *d*. The drawing illustrates how the length of all the fibres of that neuron can be calculated from the projections of the cut fibres. For explanation see text ("correction III")

drites was measured by Staiger (1984). To go from the projection to the real length of the fibres we relied on a different geometric reasoning illustrated in Fig. 33.

III. If a neuron with several straight fibre processes of length r is contained in a section of thickness d, $(d < r)$, the processes, which we suppose oriented at all angles, are cut at different points by the section. The projections of the segments on the plane of the section, if the centre of the neuron is in the middle of the section, have a length $(d/2)\tan\alpha$, depending on the angle α which they form with the normal to the section. Now we may ask how much longer the projection of each fibre would have been if the section was thick enough to contain them all, i.e. if $d > 2r$? In that case, as we know already, the projection is foreshortened by a factor $\sin\alpha$. Thus the projection of a piece contained in the histological section is to the projection of the entire fibre as $(d/2)\tan\alpha$ is to $r\sin\alpha$, which is $(d/2r)\cos\alpha$. The inverse of this, averaged over all angles α between 0 and $\pi/2$, equals $4r/d\pi$. This is the correcting factor which allows to go from the average projection of the pieces in the section to the average projection of the fibres if they were not cut. We know already that the average projection is to the length of the corresponding fibre as 2 is to π. Thus the average length of the fibres can be obtained from the average of the projections of their cut pieces in the section by the factor $\pi/2 \cdot 4r/d\pi = 2r/d$.

The surprising linear relations which result from the corrections I, IIa, IIb and III make it possible to calculate approximately the length of the fibres from their (cut) projections even if they are not all of the same length r, provided that the different lengths of the fibres are evenly represented and provided that they are all long compared to the thickness of the section.

It is comforting that method III gives almost the same result as the two corrections I and IIa if applied in tandem. In one case the measured lengths have to be augmented by a factor $2r/d$, in the other case by $\pi/2 \cdot 4/3 \cdot r/d = \pi/3 \cdot 2r/d$, which is 4% more. The difference is small compared to the uncertainty inherent in the measurements.

The choice between the alternative geometries underlying the corrections IIa and IIb (Fig. 32) is more important. If we postulate axial symmetry, as seems appropriate in the case of pyramidal cells, the value we obtain is by 18% higher than in the case of spherical symmetry ($\pi^2/4 \cdot r/d$ vs $2\pi/3 \cdot r/d$).

For convenience we summarize the results. In order to estimate the length of the fibres on the basis of their projections in camera lucida drawings, taking into account the fact that they may have been cut by the microtome, we multiply the average length measured on the tracings by a factor of

$$2 \cdot r/d \qquad \text{(correction III)}$$
$$\pi/2 \cdot 4/3 \cdot r/d = 2.09 \cdot r/d \qquad \text{(correction I combined with IIa)}$$
$$\pi/2 \cdot \pi/2 \cdot r/d = 2.46 \cdot r/d \qquad \text{(correction I combined with IIb),}$$

where r is the length of the longest fibre measured and d the thickness of the preparation. If r is small compared to d, (i.e. if the fibres are completely contained in the preparation) we only apply correction I.

Chapter 17

Length of the Axonal Ramification of Individual Neurons

The neurons of Figs. 27, 28 and 29 had axons measuring 2.7, 4.7 and 6 mm (pyramidal cells), 17 mm and 10.2 mm (stellate cells), 3.2, 4.4, and 2.3 mm (Martinotti cells) (Braitenberg, 1978b). If we apply correction IIb rather than IIa, the figures for the pyramidal cells become 3.2, 5.5 and 7.0 mm, and those for the Martinotti cells 3.8, 5.2 and 2.7 mm. These figures deserve some comments.

In all cells the measured length of the axonal ramification was less than the length one obtains by dividing the total axonal population of the cortex by the number of neurons (Chap. 8). Since we do not believe that the fibres of extracortical origin contribute very much to the total axonal population, (Peters and Feldman, 1976, showed this for the case of thalamo-cortical afferents even in the layer where they mostly terminate) we are forced to invoke other factors which may explain why the measurements of the axons of individual neurons fall short of the extent of their real axonal ramification. One obvious explanation is that the neurons may be incompletely stained and this is indeed suggested by the comparison of Golgi material with neurons stained by intracellular injection (e.g. Parnavelas et al., 1983); another, that only a small part of their axon is contained in the section. This is certainly true for the pyramidal cells, since we know from Cajal and others that most of them have a long axon leaving the cortex and re-entering it in a different place in order to produce a terminal ramification which may be at least as large as the collaterals in the vicinity of the cell body. These distant ramifications have never been stained in our preparations, or rather, their connection with the parent neuron could not be demonstrated. Thus the contribution of pyramidal cells to the intracortical axonal feltwork is likely to be twice as large as our measurements of the local ramification indicated. But this is still almost an order of magnitude less than the expected length of 4 cm per neuron.

Also the Martinotti cells may have much of their axonal tree outside of the histological section, in a way which is not corrected for by the methods outlined in the previous chapter. Many of them reach the first layer of the cortex and produce long collaterals there which can hardly be followed very far.

The stellate cell with its typically local axonal tree is less affected by this, and indeed the measured length in their case is only by a factor 2 or 3 less than the expected average length.

There was the suspicion that the small number of neurons measured may not be a representative sample. Measurements were therefore performed by Staiger on 58 pyramidal cells with well-stained axons and on 11 non-pyramidal cells (Staiger, 1984). Figure 34 shows the distribution of the axonal length in the pyramidal cells, ranging from less than 1 mm to almost 17 mm. The calculation was according to method III of the preceding discussion (Chap. 16). The neurons with fairly large axonal trees were quite rare, the majority being close to the neurons of Fig. 27.

One observation on this sample is perhaps relevant to our puzzle. The (uncorrected) values obtained by Staiger when measuring the length of the projection of the axonal tree on the camera-lucida tracing varied much less in the sample than the length calculated according to the methods IIa, IIb and III. This is because the corrections depend on the length of the longest fibre observed, and this varied a great deal from neuron to neuron, simply because it depends on chance whether a long fibre is completely present in the histological section or whether it leaves the section because of its oblique orientation. This variation is not considered in the methods we used, and it may severely affect the results, especially when the axonal tree is composed of relatively few and relatively long collaterals, as is the case with (at least some) pyramidal cells (Fig. 35).

Figure 36a and b illustrates this point. Consider an axon collateral with a length of 300 μm. It may be oriented at any angle α, with respect to the plane of the section, which is 80 μm thick (the average thickness in our sample). The neuron is located in the middle of the section. The length measured is $l = 40\ \mu\text{m} \cdot \tan \alpha$, but does not exceed 300 μm. With what probability do we measure a length between l and $l + \Delta l$ (Δl being

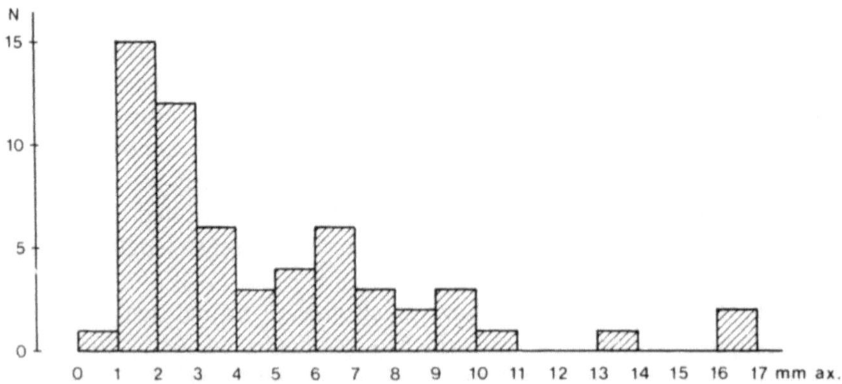

Fig. 34. Distribution of the axonal lengths (of the local ramification) of 58 pyramidal cells in the mouse cortex. The length varies from less than 1 to 17 mm. The calculation was according to correction III. This distribution should be compared to Fig. 36a and b, which provides a possible explanation for the preponderance of the smaller values of axonal length in the observed sample. *ax* axonal length of individual neurons in mm; *N* number of neurons

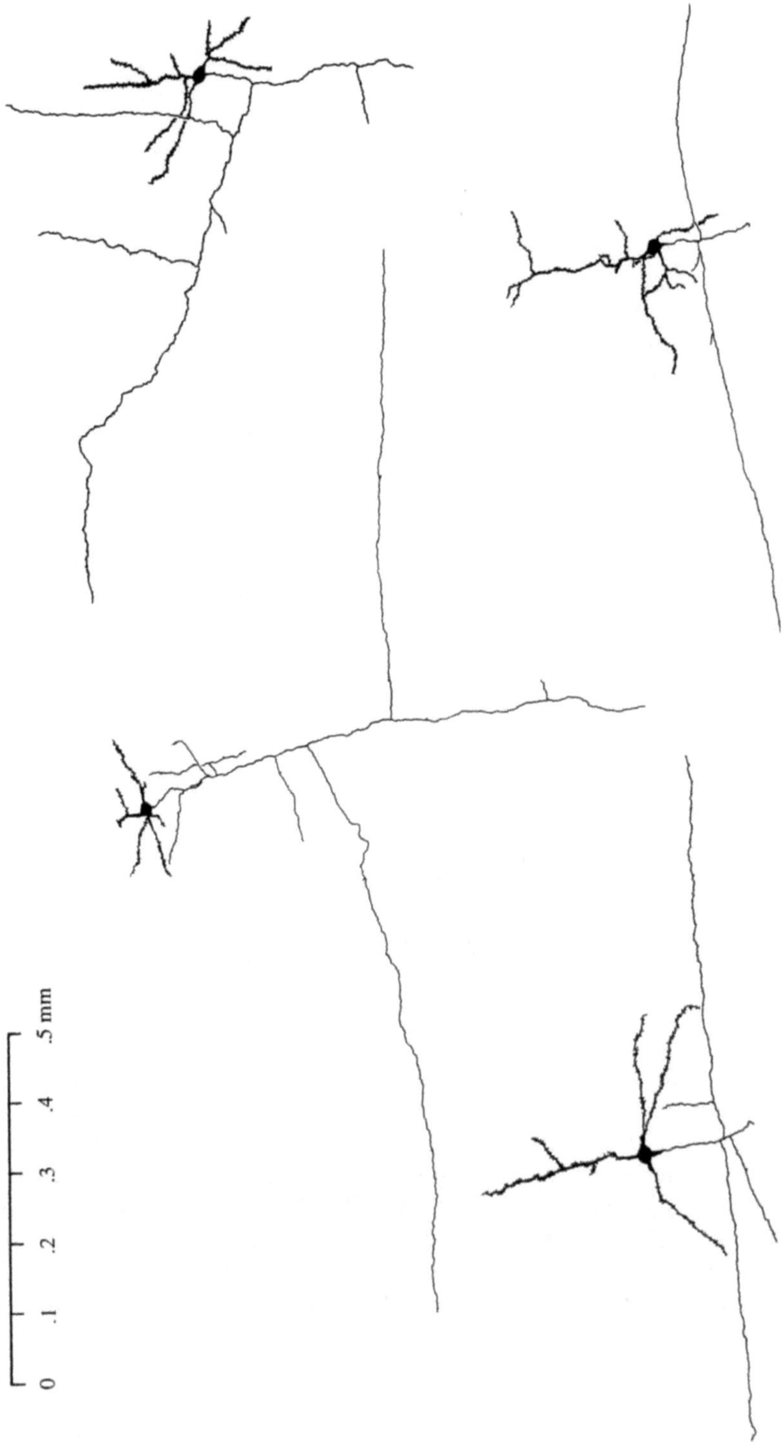

Fig. 35. Pyramidal cells of the mouse cortex with long axon collaterals, measuring several hundred microns. Note the much smaller magnification of Fig. 35 compared to Figs. 27 and 28. The thickness of axons and their collaterals is exaggerated in the drawing

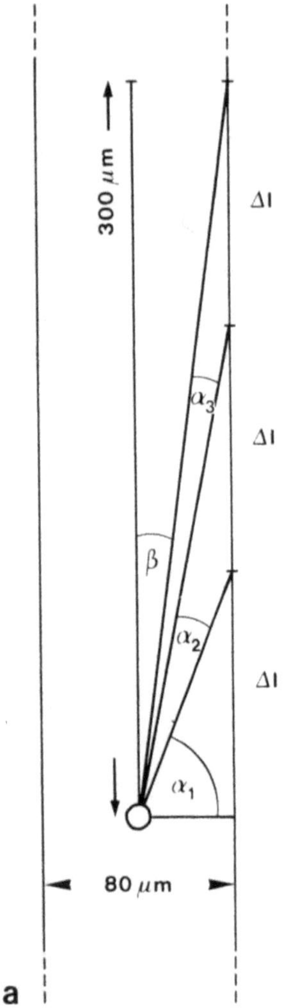

Fig. 36 a.
If in a histological preparation fibres of all orientations occur with the same frequency, the probabilities of their projections having length 0–Δl, Δl–$2\Delta l$, ... are proportionate to the angles α_1, α_2 ... The expected distribution is shown in **b**

the bin width in the histograms of Fig. 34 and 36b)? The probability is proportionate to $\alpha_{n+1} - \alpha_n = \arctan(l + \Delta l) - \arctan l$, and this varies a great deal with l (Fig. 36b). Since the total length of the axonal tree calculated from the length of its projection according to our methods IIa, IIb, III depends linearly on the length of the longest fibre measured, we may suppose that the variation shown in Fig. 34 is reducible to that constructed in Fig. 36b. In other words, we are entitled to assume that all axonal trees are in reality of the same length, corresponding to the longest measured. The other measurements would then be artefacts due to the longest collateral being severed in the section.

Thus with our longest axonal trees, measuring 17 mm or so, now becoming representative for the rest of them, we are coming close to the

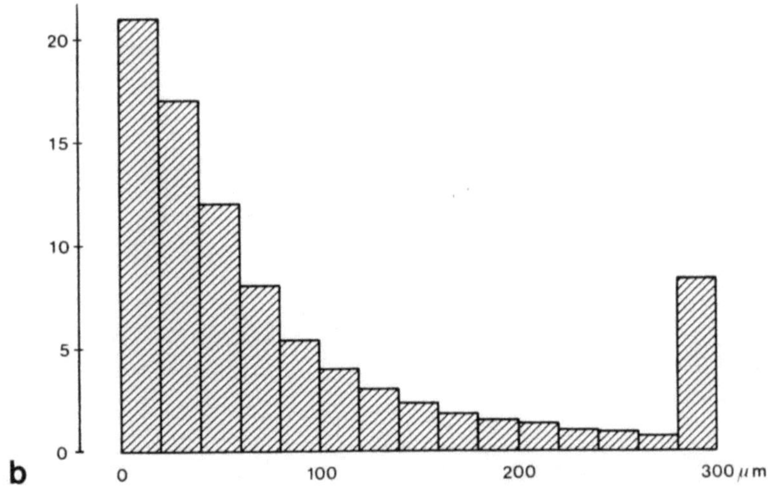

Fig. 36 b. If the fibres are of limited length, in our example 300 μm, there will be a peak in the probability of the longest projections, since the angle β, representing the probability of all fibres not cut by the section, will be added to the last α_n. This geometry provides a possible explanation for the distribution of axonal length shown in Fig. 34, under the supposition that in reality all neurons have long axon collaterals, which are but rarely contained in the section in their entirety

puzzling figure of 40 mm of intracortical axon per neuron which we had obtained by our global measurements. If we double the value obtained for pyramidal cells, by considering their distant cortico-cortical terminations which could not be measured in Golgi preparations, the discrepancy is almost entirely resolved. We may safely assume something like 20 to 40 mm of intracortical axon for each cortical neuron.

The two "stellate" cells of Fig. 28 also deserve comment. Whereas the one on the left represents very well the stellate type as we have defined it in Chapter 15, the other has traits which could justify its inclusion among the Martinotti cells. Its axon leaves the cell body in a clearly upward direction and there is a definitive difference between the main axon and its collaterals. Also, the appearance of the axonal branches is more "scraggly" than in the cell on the left. The size, the general shape of the axonal and dendritic trees and the overlap of the two are, however, quite similar in the two cells.

Chapter 18

Length of the Dendritic Trees

We measured dendrites with the same methods we had used for axons. Here again, besides the few specimens measured in Braitenberg (1978b), we relied on Staiger's (1984) collection. The sample was larger than in the case of the axons, since satisfactory impregnation of the dendritic tree is much more common in Golgi preparations than a complete stain of both dendrites and axons. The stain of the initial segment of the axon, in conjunction with the shape of the dendritic tree, is sufficient in most cases to confidently classify the neurons as belonging to one of our three types.

For the neurons measured by Staiger (see Fig. 30 for their localization) the average length of the dendrites was 3.08 mm for pyramidal cells, 2.16

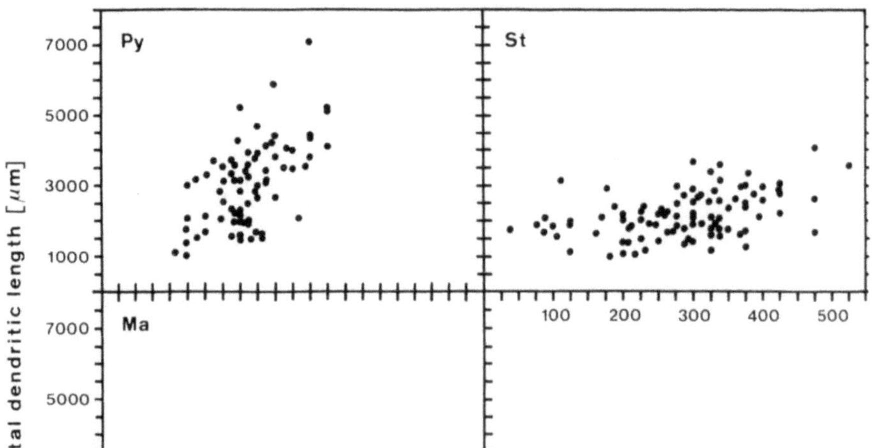

Fig. 37. Each *dot* on the three scatter diagrams represents one neuron. The *abscissa* indicates the width of the dendritic tree (its maximum lateral extension), the *ordinate* the total dendritic length. The scatter suggests a linear relation for pyramidal cells *(Py)*. In the case of stellate *(St)* and Martinotti cells *(Ma)* the width varies a great deal, while the total dendritic length remains fairly constant. It is less in the Martinotti cells than in the stellate cells

mm for stellate cells and 1.47 mm for Martinotti cells. Staiger only applied correction I to his data, which may be one of the reasons why his figures are smaller than those in our first report (pyramidal: 3.1, 3.2, 4.0; stellate 4.5, 6.7; Martinotti: 2.3, 3.2 and 4.4 mm). However, they are probably not far from the correct value. In Chapter 11 we had noted that the density of dendrites and axons measured on electron micrographs, as well as the spacing of synapses on dendrites and axons, suggests an average length of the axonal tree almost ten times that of the dendritic tree.

Staiger's measurements provide additional confirmation of our classification of cortical neurons in at least two distinct types. Figure 37 plots for each neuron the total length of dendrites against the lateral spread of the dendritic tree. The class of pyramidal cells is distinct from the other two in that it shows a rough proportionality of the two measures. On the contrary, in both stellate and Martinotti cells the total dendritic length seems to be quite independent of the width of the dendritic tree, the two classes being, however, distinguished by the greater average dendritic length of thc stellate cells. This seems to indicate that pyramidal cells come in different sizes, but have fairly uniform shapes, while the non-pyramidal cells tend to have, within each class, a constant dendritic length but dendritic arbors more or less deformed, the lateral spread varying a great deal.

Chapter 19

Relative Density of Axons and Dendrites

We found two questions interesting in connection with the shape of neuronal ramification in the cortex. One is, how wide is the territory in which the intracortical axonal branches of a certain neuron are distributed? The other, what proportion of the synapses present within that territory is served by that one axon? The two questions are related, and the answer to one can be deduced from the answer to the other once we know, as we already do, the total density of axons in the tissue and the axonal length of individual neurons.

As *relative axonal density* we define the ratio between the axonal length of one neuron and the total length of all axons present in the territory of its ramification. The latter is, in practice, the volume of a simple geometric body which approximates that territory, times the overall axonal density as considered in Chapter 7. For example, a pyramidal cell with a (local) axonal tree of 20 mm, distributed in a sphere of radius 0.3 mm (the length of the longest collateral), assuming a density of 4 km of axon per mm^3, has a relative axonal density of 4×10^{-5}. Supposing a homogeneous distribution of synapses on all axons, we would conclude that the pyramidal cell contributes little more than one hundred-thousandth of the synapses present in the territory in which it is ramified.

Take one of the large stellate cells of the kind illustrated in Fig. 27. Let us again assume 20 mm of axonal length but this time distributed within a sphere of radius 0.15 mm. The relative axonal density turns out to be 3×10^{-4}, almost an order of magnitude greater.

Only in the central part of the axonal ramification of stellate cells, or in some terminal ramifications of afferent fibres (Fig. 31), does the relative axonal density come up to the order 10^{-3}. We have not seen an axonal ramification of higher density than that in the mouse cortex.

This leads to an important proposition: nowhere in the cortex does any one neuron dominate more than about one thousandth of the synapses within the territory of its axonal spread. *No one element has the absolute say, anywhere.* Every signal, be it of external or intracortical origin, is intimately mixed with the signals from many different sources.

Unless, of course, we postulate a high specificity of intracortical neuronal connections, which may have escaped our statistical view of cortical connectivity. We will come back to this point in Chapter 21.

Relative dendritic density can be determined in a similar way. We were satisfied with a rough estimate of the order of magnitude, which is 10^{-3}

for the basal dendrites of pyramidal cells. This means that a certain dendritic spine, chosen at random, has a likelihood of only 0.001 of belonging to a given pyramidal cell whose basal dendrites reach that region. We will use this figure in the following chapter in order to derive the probability of connections between cortical neurons.

Chapter 20

The Likelihood of a Synapse Between Two Pyramidal Cells

We are now in a position to estimate the probability of one, two or more synapses between the axonal tree of one neuron and the dendrites of another given neuron situated at a certain distance from the first. We shall limit ourselves to synapses between pyramidal cells which are the most common synapses in the cortex and will disregard the distant connections through cortico-cortical fibres, about whose geometry we know very little. Thus, what we have in mind are the local connections between the axon collaterals of one pyramidal cell and the basal dendrites of another.

We used two different arguments, one purely statistical and the other geometric. The quantitative results are similar.

I. We may stylize the receiving neuron as a cloud of postsynaptic sites, represented by the spines on its basal dendrites. There are several thousands of them within a roughly spherical region, and we suppose that their location within that region is random. Because of the overlap with the dendritic trees of neighbouring neurons, the postsynaptic sites belonging to the neuron in question are intermingled with 1000 times more postsynaptic sites belonging to the other neurons (Chap. 19), altogether several millions (there are, according to our analysis in Chap. 5, 10^6 synapses in a cube of cortical grey substance with side of 100 μm, and about four times more in a sphere of that radius).

Now we imagine an axon collateral of a nearby neuron entering the dendritic tree of our neuron. The collateral establishes synapses spaced at an average distance of 5 μm (Chaps. 8 and 9), picking the postsynaptic elements at random (Chap. 10). If the collateral is 100 μm long, it will make 20 synapses. The probability for any one synapse to have the particular dendritic tree as the postsynaptic partner is $p = 0.001$. What is the probability $(w_0, w_1, w_2, \ldots)$ that among the $n = 20$ contacts there are $k = 0, 1, 2\ldots$ synapses with that particular neuron?

The binomial distribution

$$w_k(n,p) = \binom{n}{k} p^k (1-p)^{n-k}$$

provides the answer (Fig. 38): the most likely event is that no contact is made ($w_0 = 0.98$), the probability of one contact is quite small ($w_1 = 0.019$) and the probabilities for more than one contact are negligible ($w_2 = 0.00018, w_3 = 0.000001$). Figure 38 also shows the probabilities

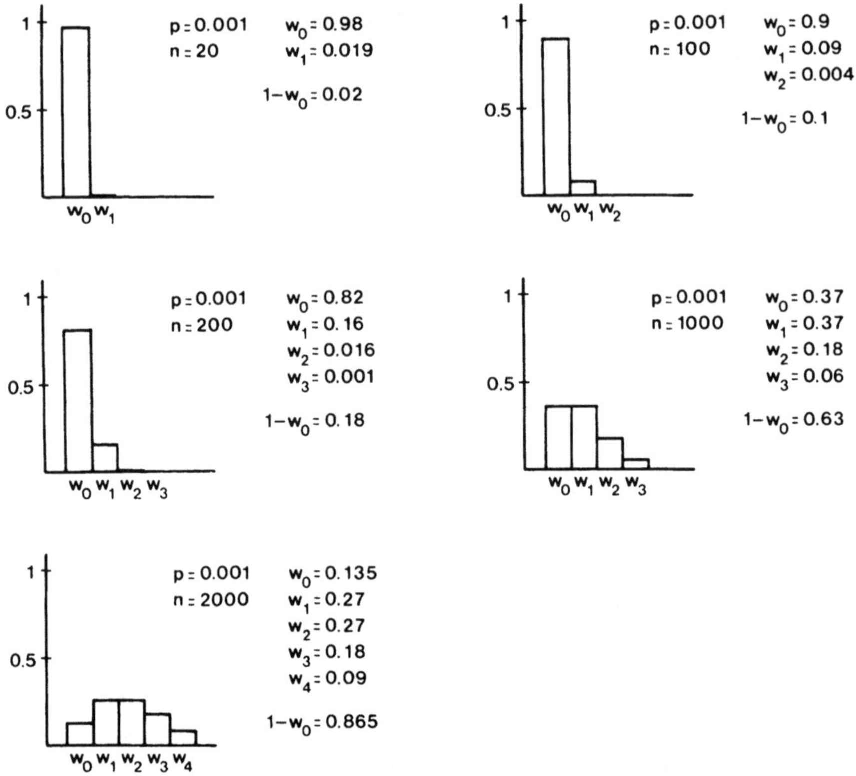

Fig. 38. Probability of connection between two cortical cells, under the assumption that clouds of presynaptic sites mix with clouds of postsynaptic sites, the connections being formed by chance. The probability of hitting a postsynaptic site of one particular neuron (whose dendrites are intermingled with the dendrites of many others) is $p=0.001$. If an axon penetrating that dendritic tree makes 20, 100, ... 2000 contacts, the probabilities w_0, w_1, w_2 ... of establishing no contact, one, two etc. contacts with a particular neuron can be calculated. It can be seen that even with intimately mixed dendritic and axonal trees ($n=1000$, $n=2000$) the probabilities for multiple contacts are quite small. (Braitenberg, 1978 a)

in the case of axons making $n = 100, 200, 1\,000$ and $2\,000$ synapses. They correspond to a more or less intimate mixing of the axonal tree of one neuron with the dendritic tree of the other. It can be seen that only for the most favourable cases corresponding to neurons at very close distance do the probabilities for two or more contacts emerge. On the whole, in most cases we must suppose that two neurons, with the axonal tree of one overlapping the dendritic tree of the other, make no synaptic contact or just one synapse.

II. The other argument is based on the fact that the axon collaterals of pyramidal cells, as well as their secondary branches, tend to take a remarkably straight course through the tissue (Fig. 39). They "shoot"

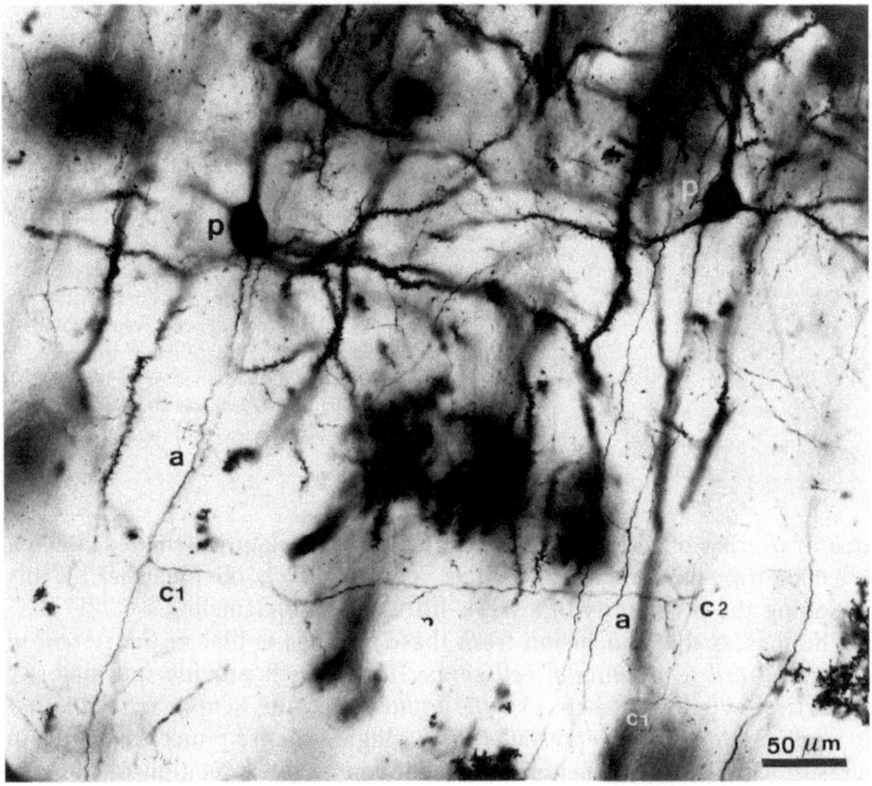

Fig. 39. Golgi preparation of the mouse cortex. The cell bodies of two pyramidal cells are marked *p*. Their descending axons *(a)* have collaterals of first *(C1)* and second *(C2)* order, all taking a remarkably straight course in different directions. The same point is illustrated in Fig. 25 and Fig. 35

through the dendritic trees of other pyramidal cells as a bullet shoots through the crown of a tree, and this simile provides the basis for an estimate of the probabilities for a single or for multiple hits. What is the probability of a bullet hitting a branch of the tree, provided it enters its crown? The probability is proportionate to the "cross-section" of the tree, to use a term from particle physics, which is the area of the projection of the tree parallel to the shooting line, divided by the area of the projection of the entire crown. Figure 40 shows the "cross-section" of the basal dendritic tree of a pyramidal cell. The thickness of the dendrite has been augmented by twice the length of the spines, since that is the region within which an axon is likely to make a synaptic contact. The area thus defined is 0.11 times the area of the circumscribed circle, and the probability of hitting it at least once is therefore $w_{1,2\ldots} = 0.11$. The probability of missing it is $w_0 = 0.89$. The picture also provides an estimate of the probability w_2 of hitting the dendritic tree twice. This is given by the

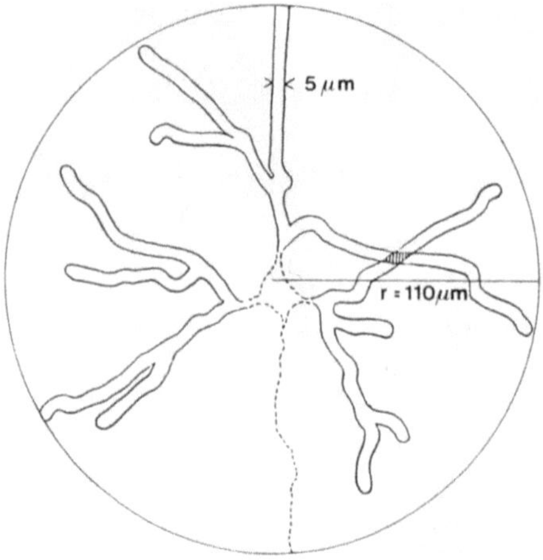

Fig. 40.
The "cross-section" of a dendritic tree (basal dendrites only) of a pyramidal cell. This is the projection of the dendritic tree on a plane, the thickness of the dendrites being augmented so as to include their spines. The area of the projection occupies 11% of the area of the circumscribed circle. The area of overlap of two dendrites *(shaded)* is 0.06% of the total. These areas provide an estimate of probability of hitting that dendritic tree once or twice. See, however, Fig. 47, which suggests even lower probabilities. (Braitenberg, 1978a)

area of overlap of different dendrites of the same neuron (shaded), which is 0.0006 times the area of the circle. It is intuitively obvious that by this reasoning the probability of a triple hit is negligibly small.

The inescapable conclusion from these findings is that *in the system of pyramidal cell to pyramidal cell connections,* which provide the majority of cortical synapses (Chap. 15), *the influence of one neuron onto another is weak,* being mostly represented by a single synapse. Since this is again a result which is fundamental to our functional interpretation of the cortex, we have to dwell on it and examine its implications.

If multiple synapses between the same two neurons are rare, it follows that each cortical pyramidal cell *contacts almost as many other pyramidal cells as it has synapses,* which is, on its local axonal arborization, about 4000. These are distributed in a region perhaps 1 mm across (twice the length of the longest axon collaterals plus twice the length of long basal dendrites). Within this region, even quite close to the pyramidal cell in question, there are many other cells which are not contacted by it.

The utmost divergence of signals from one cell to thousands of others corresponds, of course, to a similar convergence: the thousands of synapses on the dendritic tree of a pyramidal cell are contacted by axons from about as many different pyramidal cells.

If our premises up to this point are correct, a picture of the cortex emerges as a device for the most widespread diffusion and most intimate mixing of signals which is compatible with the natural limitations of nerve cells. It is important to realize that this property, if it is not unique to the cerebral cortex, is certainly characteristic for it. There are many other pieces of nerve tissue where the connections are much more specific: the cerebellum, the lateral geniculate body, the visual ganglia of insects, etc.

The loose, weak and diffuse coupling is characteristic for the system of pyramidal cells. We like to call this the *skeleton cortex*, because pyramidal cells are the most numerous cell type and especially because the large majority of synapses in the cortex are synapses between pyramidal cells. For the minority of synapses which do not belong to this system, including the type II synapses which have non-pyramidal cells as their presynaptic element, the statistics of the connections may be very different. The so-called chandelier cells (for the terminology see Fairén et al., 1984) make inhibitory synapses onto the initial segment of pyramidal cell axons (Somogyi, 1977), and other cells of the stellate kind may have a special affinity with the cell body or with the dendritic shafts of other neurons, or produce such dense axonal arbors that we must assume multiple contacts to be quite common.

We shall complete this chapter by comparing the results in Braitenberg (1978b) and in Braitenberg and Schüz (1991) to more recent theoretical and experimental works.

Liley and Wright (1994) calculated the probability of connection with a different mathematical approach. Among other things, they used clouds of straight axonal and dendritic segments, rather than clouds of pre- and postsynaptic sites. The results are very similar to ours.

Hellwig (1994; 1995, and in preparation) followed an approach in which he used real ramification patterns of neurons. He did this by injecting biocytin into neurons of layer II/III in slices from the visual cortex of the rat. He then drew the axonal and dendritic trees of these neurons by way of a Camera lucida mounted onto a computer microscope, so that the 3-dimensional structure of the arborizations was maintained in the reconstructions. He then made all possible pairings between the axonal and dendritic trees of his collection and, by way of a computer program, shifted the axonal tree and the dendritic tree of each pair stepwise against each other in a tangential direction. Taking steps of one micron, all possible contacts were recorded. A contact was assumed if a piece of dendrite and a piece of axon fell into the same cube of 1 micron side length. The results match ours well, the probabilities of a connection between a given pair of neurons ranging from 0.8 to less than 0.1, and, in addition, give detailed information as to how these probabilities depend on the distance between the neurons. The most favourable case, corresponding to the lowest diagram in Fig. 38 (with a connection probability of about 0.8) only occurs when the neurons are no more than 50 μm away from each other. Within this range, about 33% of the neurons make 2 or 3 contacts with each other, and about 7 % make multiple contacts of 6 or more. Even in close vicinity, the most frequent case is one contact (about 23% of the neurons up to a distance of 50 μm). The fall-off of the probability with distance depended somewhat on the relative depth of the neurons. When both neurons were located in layer III, a probability of 0.5 was reached at a distance of 150 μm, a probability of 0.1 at about 350 μm. This fall-off is probably somewhat steeper in the mouse, since the rat can be assumed to have wider axonal and dendritic trees than the mouse (see chapter 34), although the fall-off is probably not quite as steep as we had assumed in the previous edition of this book.

Lübke et al. (1996) also investigated intracellulary stained neurons. They searched for autapses (=synapses between an axon and a dendrite of the same neuron) on intracellularly stained layer V pyramidal cells. In 80% of the neurons, the axon approached the dendrites so closely that an autapse could be assumed (and could sometimes be confirmed with the electron microscope). This corresponds well to the calculations above, since the formation of autapses corresponds to the most favourable case assumed in Fig. 38 and to a zero-shift in Hellwig's experiments.

Electrophysiological results are, on the whole, also in good agreement with the calculations, even though they suggest lower probabilities of connection. Using pairs of electrodes, Mason et al. (1991) searched for monosynaptically connected neurons. Experiments were carried out in layer II/III of slices of the rat cortex. The probability of finding a neuron presynaptic to another given neuron, in a population of neurons 50 to 340 μm apart from each other (average 140 micron) was 0.09. This is somewhat lower than Hellwig's data would predict for this range of distances. Even lower probabilities (about 0.04) were found in similar experiments by Thomson et al. (1988). In this case, the presynaptic neurons were located deeper than the postsynaptic neurons, in a fanshaped region 50 to 300 μm below and up to 100 μm lateral or medial from the postsynaptic neuron.

The fact that electrophysiological data tend towards lower overall probabilities of connection than the anatomical ones is not surprizing. The "hits" between axons and dendrites in the anatomical calculations give only an upper limit of the possible number of synapses. On the other hand, electrophysiological results probably give a lower limit for the number of connected pairs of neurons, since connections through one synapse, which are the majority, may escape electrophysiological detection. This may also explain the relatively high frequency of multiple synaptic contacts in the pairs of neurons using the electrophysiological technique. The multiple contacts could, however, also be due to the fact that these recordings were made in layers V/VI, where pyramidal cells have particularly large dendritic trees which increase the probability of hits when they act as the postsynaptic partner. Markram et al. (1997) found a pair of neurons with 5 synapses in one direction and 6 in the other. Deuchars and Thomson (1995) found 3 synapses in one case, and 6 in another, with 6 additional synapses suspected but not confirmed. In all these cases, the neurons were very close to each other or only separated in depth. Lübke et al. (1996) found an average of 5.6 possible contacts in 13 pairs of neurons in layer V, but this is again an upper limit, since the light microscope was used.

Alltogether, our view of the skeleton cortex as an ensemble of weakly connected pyramidal cells with a high degree of divergence and convergence, finds ample confirmation.

Chapter 21

Peters' Rule and White's Exceptions

The overall statistics of cortical elements which we had developed (Chaps. 3–12) even before the concept of neuronal types was introduced (Chaps. 14, 15) has not been seriously challenged by any of our later more detailed measurements. If we imagine a cortex in which pyramidal cells and non-pyramidal (or briefly, stellate) cells are thoroughly mixed, and if we suppose that both cell types carry about the same number of synapses and distribute them without specific preferences, we can roughly predict the proportions of various types of synapses in the electron microscope. With 85% pyramidal cells (*py*) and the rest stellate cells (*st*) we obtain:

py to *py* synapses:	85%×85% = 72% Type I on spines[a]
py to *st* synapses:	85%×15% = 13% Type I mainly on dendritic shafts
st to *py* synapses:	15%×85% = 13% Type II on cell bodies and dendrites
st to *st* synapses:	15%×15% = 2% Type II mainly on dendritic shafts

[a] neglecting the few Type I synapses on the dendritic shafts of pyramidal cells.

These predictions are not far from the truth. On electron micrographs we counted 75% synapses residing on spines, only slightly more than predicted by the table. On the same material, we estimated 11% Type II synapses, somewhat less than the 15% one would expect if pyramidal cells and stellate cells really produced the same number of synapses. Since (Chap. 9) according to Hellwig (1992) the density of synapses along the axon is about the same in the two types of neurons, both discrepancies may mean that the axons of pyramidal cells are longer, on average, than those of stellate cells, or alternatively, that the ratio of pyramidal to stellate cells is 87/13, rather than the estimated 85/15. In that case, supposing axons of the same length for both types of neurons, the probability of synapses on spines (= synapses between pyramidal cells) turns out as 0.76 and the probability of Type II synapses as 0.13, quite close to the measurements.

This play with probabilities is legitimate only if synapses between cortical neurons are made entirely by chance, depending only on the accident of some axon of one neuron coming into the immediate vicinity of some dendrite of another. We would then expect the distribution of synapses of various origins (e.g. type I synapses from pyramidal cells and from extracortical afferents, type II synapses from various inhibitory interneurons) on the dendritic tree of any one neuron to reflect simply the availability of those presynaptic elements in the tissue, and we would expect the proportions of synapses of different types on any dendritic tree to be the same as the proportions of the corresponding elements in its environment. Conversely, the postsynaptic partners of any axonal tree would simply reflect the distribution of the postsynaptic elements of various kinds in the layers in which it ramifies. This very strong supposition in our laboratory jargon goes by the name of Peters' rule, although the principle was stated by Peters and Feldman (1976), Peters (1979) only for the statistics of the synapses between geniculo-cortical afferents and cortical neurons in the rat visual cortex. The generalization to any sort of cortical synapses is ours, with apologies to Drs. Peters and Feldman.

The random connectivity between thalamo-cortical afferents and cortical neurons has not remained unchallenged. E.L. White and his coworkers (White, 1978, 1979, 1981, 1986, 1987; White et al., 1984; White and Hersch, 1981, 1982; White and Rock, 1979, 1981) used this very same thalamo-cortical junction (although in a different area) to provide examples of a selectivity of synaptic connections which, in their opinion, goes well beyond what could be expected simply from the geometry of the distribution of dendrites and axons.

They do agree with previous descriptions to the extent that axons entering the cortex make some synaptic contacts with every sort of potentially postsynaptic element available within the territory of their terminal arborization. But, if the frequency with which the different kinds of elements are contacted are established by careful quantitative electron microscopy, they are sometimes found to be different from the frequency of these elements in the tissue. So, some special affinities must be postulated. For instance, in a certain region, (layer IV of the mouse somatosensory cortex) spiny neurons, i.e. pyramidal cells in our terminology, receive 20% of their asymmetric (type I in our terminology) synapses from the thalamocortical afferents, while in the same region a certain kind of non-spiny neurons receive less than 4% from the same source. Some stellate cells seem to repel the presynaptic offering coming from the thalamic afferents, and pyramidal cells perhaps attract it, although the 20% of thalamic afferents which they receive may simply reflect the density of the terminals from that source in the neuropil of Layer IV. The preferential contacts depend on affinities that are not quite understood. The proportion of thalamo-cortical afferents on the dendrites of spiny (pyramidal or "spiny stellate") cells was found to vary between 10 and 23%, while in the case of the non-spiny cells the range was even larger, between 0 and 21%.

The White group (Porter and White, 1986) found evidence for some synaptic selectivity also outside the thalamocortical link, e.g. in the synapses between the neurons projecting through the corpus callosum and callosal neurons of the other hemisphere projecting in the opposite direction. Dendrites of callosal neurons residing deep in the cortex receive only half the number of synapses from their contralateral partners which the dendrites of other callosal neurons residing in the upper tiers of the cortex receive, even when they are intermingled in the same layer and are served by the same incoming fibres. On the other hand, in various studies of the synapses established by axon collaterals of pyramidal cells with neurons of the same hemisphere (Kisvarday et al., 1986; White, 1989), over 80% of synapses resided on spines of other neurons, quite in accordance with the proportion of spine synapses among all Type I synapses. This is not surprising and confirms Peters' rule. In other cases (McGuire et al., 1984; Winfield et al., 1981), less than half of the synapses were reported to reside on spines, suggesting an active avoidance of pyramidal cells as the postsynaptic partners on the side of these particular neurons.

The most impressive exception to Peters' rule was described by White and Keller (1987) in the case of the local axon collaterals of cortical cells projecting from the somatosensory area of the mouse to the thalamus. These have the appearance of ordinary pyramidal cells, and indeed the synapses formed by their axons are of Type I. However, collaterals which these axons give off in the cortex before leaving for the thalamus, instead of contacting the spines on the dendrites of other pyramidal cells, as most other pyramidal axon collaterals do, establish contacts almost exclusively with the shafts of spineless dendrites, presumably belonging to stellate cells.

Thus there is no doubt that Peters' principle is subject to exceptions even in the cortex, where we believe it to be generally valid, but the reports of various statistical biases in the pattern of synaptic connections up to now are not such as to invalidate the approach we have taken in our analysis.

Chapter 22

Dendritic Spines

We mentioned dendritic spines in connection with the distinction of different types of synapses in the electron microscope (Chap. 12) and then again in Chapter 15, where they served as one of the main criteria for the definition of different neuronal populations in the cortex. Some neurons have almost as many spines as they have (afferent) synapses, and this seemed important enough to define a special class, that of the pyramidal cells (or spiny neurons, as some would prefer) irrespective of the shape of the dendritic and axonal ramification. Other neurons have no spines, or very few spines; moreover, for these few spines it has never been shown that they are really the same as those of the pyramidal cells, which have a very characteristic appearance on electron micrographs.

Figure 41 shows spiny dendrites from a Golgi preparation, and Fig. 42, at higher magnification, some longitudinal sections of spines as they appear on electron micrographs. There is a spine head (*h*) and a spine neck (*n*) or stem. Electron microscopists have learned to recognize heads of spines on their sections even if their attachment to the dendrite is not evident, because of the "fluffy" appearance of the cytoplasm within. The neck contains always a peculiar membrane system, perhaps a folded cistern or a group of cisterns called "spine apparatus" (*a*). There is always a synapse of Type I occupying part of the surface of the head (*s*), and occasionally, but quite rarely another synapse which is then always Type II[1] (Colonnier, 1968).

In spite of the constant features which characterize spines on electron micrographs, their size and shape and also their orientation with respect to that of the parent dendrite vary a great deal, as can be seen in the Golgi picture.

Spines are found on dendrites in various places of the nervous system, not only in the cerebral cortex, and in some cases, e.g. on the Purkinje cells of the cerebellum, they have been shown to be quite of the same kind as in the cortex, with all the features showing in the electron microscope. In the cerebral cortex they are obviously very important, since there, as we have already said, they are associated with 75% of all synapses.

[1] In a survey of 106 spines in serial sections we found no spine without a synapse and five spines with two synapses, where the second synapse in three cases was clearly of Type II and in two cases dubious.

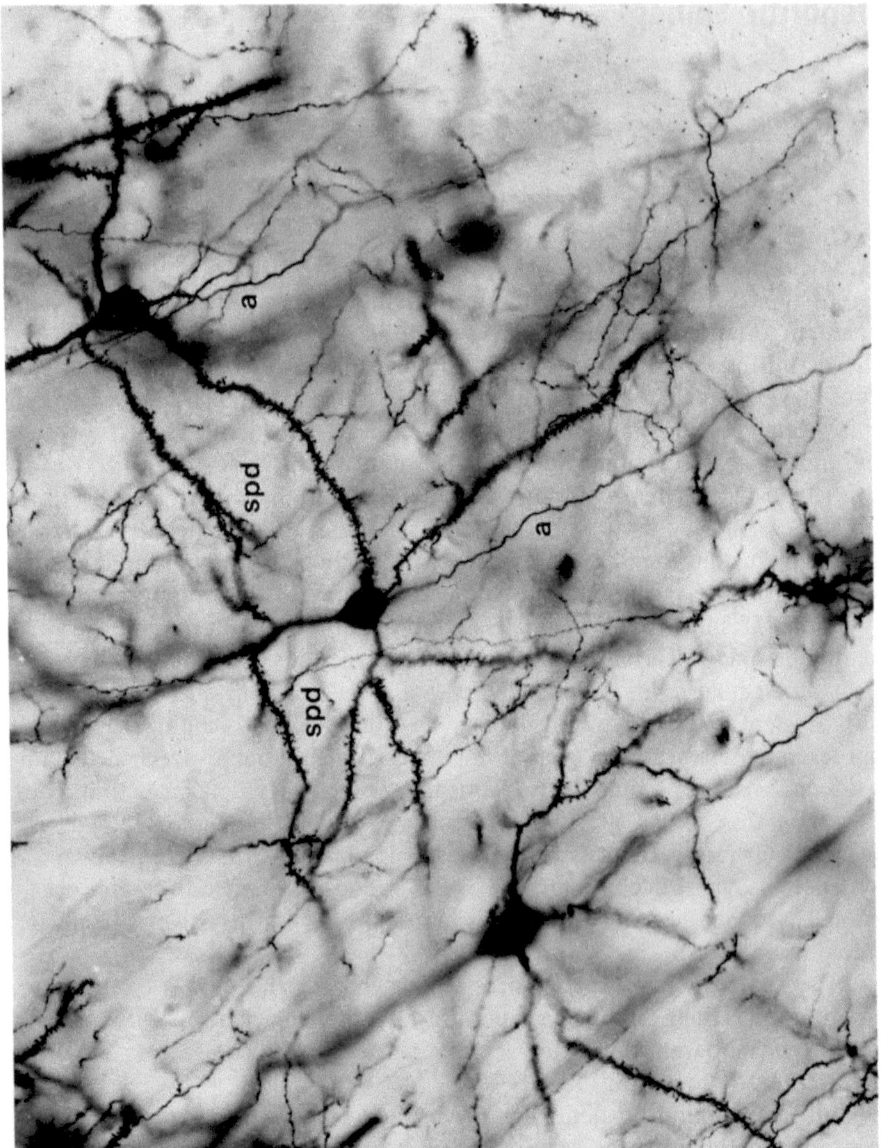

Fig. 41. The spiny dendrites *(spd)* of pyramidal cells. *a* axon

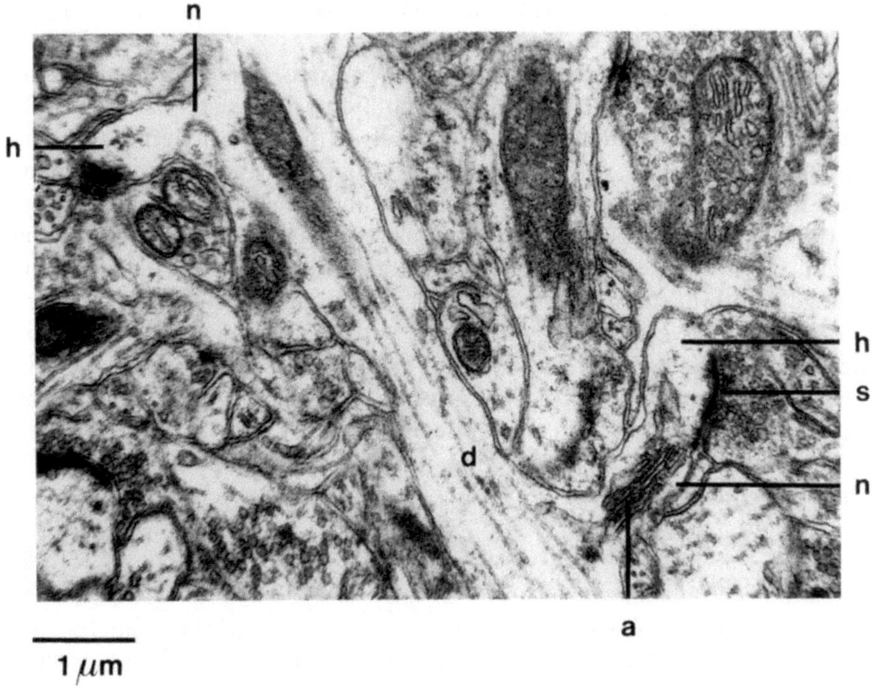

Fig. 42. Dendritic spines. Electron micrograph. *d* dendritic shaft; *s* synapse; *h* head of spine; *n* neck of spine; *a* spine apparatus

Various explanations have been proposed for these peculiar appendages of some dendrites. One of them was that the increase of dendritic surface due to the spines might accomodate a larger number of synapses. True, the surface of a dendrite is increased threefold by the presence of spines, compared to a spineless dendrite of comparable thickness. But if the number of synapses on a certain length of dendrite is counted, synapses turn out to be about as numerous on the spineless dendrites (see the following chapter) as on the spiny ones. If anything, the number of synapses is slightly higher on the spineless dendrites (Schüz and Dortenmann, 1987).

The comparison between the synaptic density of spiny and smooth dendrites implied the use of different methods. For the first, the electron microscope served only to establish that most of the synapses on pyramidal cell dendrites reside on spines (Peters and Feldman, 1977; White and Hersch, 1981). Spines (= synapses) were then counted with the light microscope in Golgi preparations.

Such counts are tricky. For one thing, spines are so small that the limit of resolution of the light microscope becomes a problem. For reasons of wave optics the photographic picture of a spine may be very different from its true shape. Moreover, spines protrude in all directions from the dendrites so that some of them are generally hidden behind or, what amounts to practically the same, in front of the heavily stained dendritic stem.

Figure 43 shows the geometry which was used to infer the true number of spines from counts in Golgi preparations. The case is fairly simple when the spines are idealized as being all of the same length, since the proportions of spines which are hidden ($2\beta/\pi$) can then be easily calculated from the thickness of the dendrites and the length of the farthest protruding spines. In the more complicated case of spines of varying length, the average length of the spines must be calculated first from the average length of the projections of the protruding spines. There is a proportionality between the two averages, with a factor which depends on the length of the spines relative to the thickness of the dendrites. This is, however, practically insignificant, since for an infinitely thin dendrite the factor is $\pi/2$ and for a very thick one 3/2 (Schüz, 1976).

The formula used for calculating the total number N of spines on a piece of dendrite was then:

$$N = \frac{n \cdot \pi/2}{\arccos \frac{r}{r+l}} \tag{10}$$

where n is the number of visible spines, r the radius of the dendrite, and l the real average length of the spines which was 1.5 μm (Schüz, 1989). For apical dendrites (with an average diameter of 1.8 μm), this yielded $N = 1.32\,n$. For basal dendrites (with an average thickness of 0.9 μm) it was $N = 1.16\,n$.

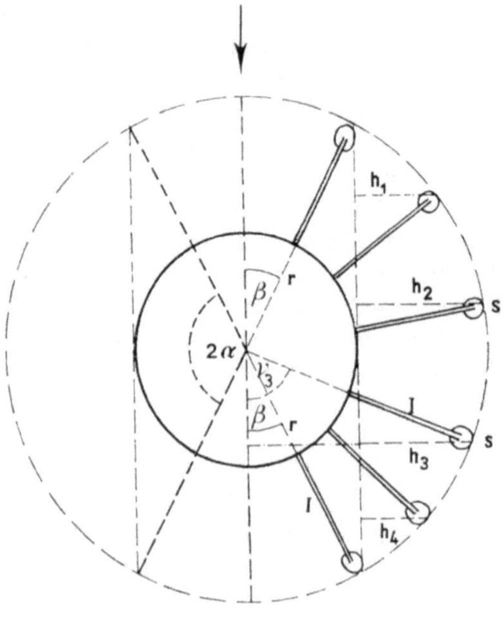

Fig. 43.
The geometry involved in calculating the true number of spines from the spines seen in Golgi preparations. Spines are invisible when they are located above or below the stem of the dendrite to which they are attached. The diagram is a projection in the direction of the long axis of the dendrite, whose cross-section is drawn as a circle. The proportion of spines hidden depends on the radius r of the dendrite and on the length l of the spines and can be calculated by straightforward trigonometry. h_1–h_4 are the apparent lengths of the spines when the dendrite is viewed in the direction of the arrow. (Schüz, 1976)

With different geometric reasonings, Feldman and Peters (1979) and Rusakov and Stewart (1995) proposed a factor slightly higher than ours. But the different geometric constructions do not seem to matter as much as the statistical fluctuations between different samples.

In the sample considered by Schüz (1976), 8 to 11 dendritic segments between 20 and 40 μm long were examined in each of ten pyramidal cells from different layers of the cortex. The grand average, corrected for hidden spines in the manner described, was two spines per micrometer. There were 2.1 spines/μm on the main stem of the apical dendrite and 1.9 on the other dendrites.

Interestingly, there are pyramidal cells which are very spiny and others which are less so. There is a strong correlation between the spininess of the apical and of the basal dendrites of one and the same cell. But sometimes two pyramidal cells which were situated very close to each other differed markedly (by a ratio of 2:1) in the number of spines on their dendrites, without any obvious difference in their morphology.

There is a curious discrepancy between the counts of Schüz (1976) and Peters and Feldman (1977) on Golgi preparations on the one hand, and those made by electron microscopists, who took upon themselves the tedious chore of reconstructing dendrites from serial sections on the other. This would seem to reveal the true number of spines without fail. However, most of these counts fall short of the counts in Golgi preparations: 1 spine/μm (Vaughan and Peters, 1973), 0.6 spines/μm (Peters and Feldman, 1977), 1.2 spines/μm (White and Hersch, 1981). But occasionally a very thick apical dendrite may carry 7.2 spines/μm (Feldman and Peters, 1979). It is well possible that statistical fluctuations in the distribution of spines become very relevant in the electron microscopists' samples, which are by necessity quite small.

Another reason for the low spine counts reported by the electron microscopists may be the occasional failure to establish the continuity of a spine with its parent dendrite on serial sections.

All told, we settle for a figure of *two spines per micrometer* on the dendrites of cortical pyramidal cells, and we take this as being only slightly less than the number of all synapses there, at least for the dendritic segments which are distant from the cell body.

Chapter 23

Synapses on Spineless Dendrites

In the case of dendrites which carry no spines and therefore receive their afferent synapses directly, such as the dendrites of the various stellate cells, counts of synapses necessarily involve the electron microscope. Reconstruction of dendrites on serial sections is very time-consuming and is limited to short segments of dendrites, with the consequent hazards involved in small sample size, as we have already noted. The task would be relatively simple if individual dendrites could be marked by some stain so as to identify the synapses belonging to that dendrite, and if the sections were thick enough to show long enough segments of dendrites.

We were able to meet the two requirements with a method which evolved from our experiments with phosphotungstic acid staining (Schüz and Dortenmann, 1987, see Chap. 5). When the fixation was not optimal, the sections stained with a modification of the method proposed by Bloom and Aghajanian (1968) often contained dark dendrites, mostly of the smooth kind characteristic of stellate cells and rarely of the spiny kind which gives away the pyramidal cell. When the phosphotungstic acid stain was successful, the background was so clear that sections 200 nm thick (more than three times the thickness of the usual electron microscopic sections) could be used. On such sections fairly long segments of dendrites can be seen (Fig. 44), long enough to perform a statistical analysis of the spacing of their synapses.

Here again problems of stereology arise. They are solved in a relatively simple way because the alignment of the synapses along smooth dendrites introduces a geometric order which can be exploited. The situation which underlies our calculation is that of a section parallel to the axis of the dendrite and containing its central part. The two pieces of membrane delimiting the dendrite are then roughly perpendicular to the plane of the section and the synapses can be easily recognized (Fig. 44 and 45). If the thickness t of the section is known as well as the radius of the synaptic junction r_s, the number n of synapses counted on a length of dendrite l yields the number N of synapses on an area F as

$$N/F = n/2l\,(t + 2r_s)\ . \qquad (11)$$

In this case it is not difficult to measure the size of the synapses and even to appreciate their rounded shape, since enough dendrites are grazed by the section so as to show the synapses from their broad side (Fig. 44).

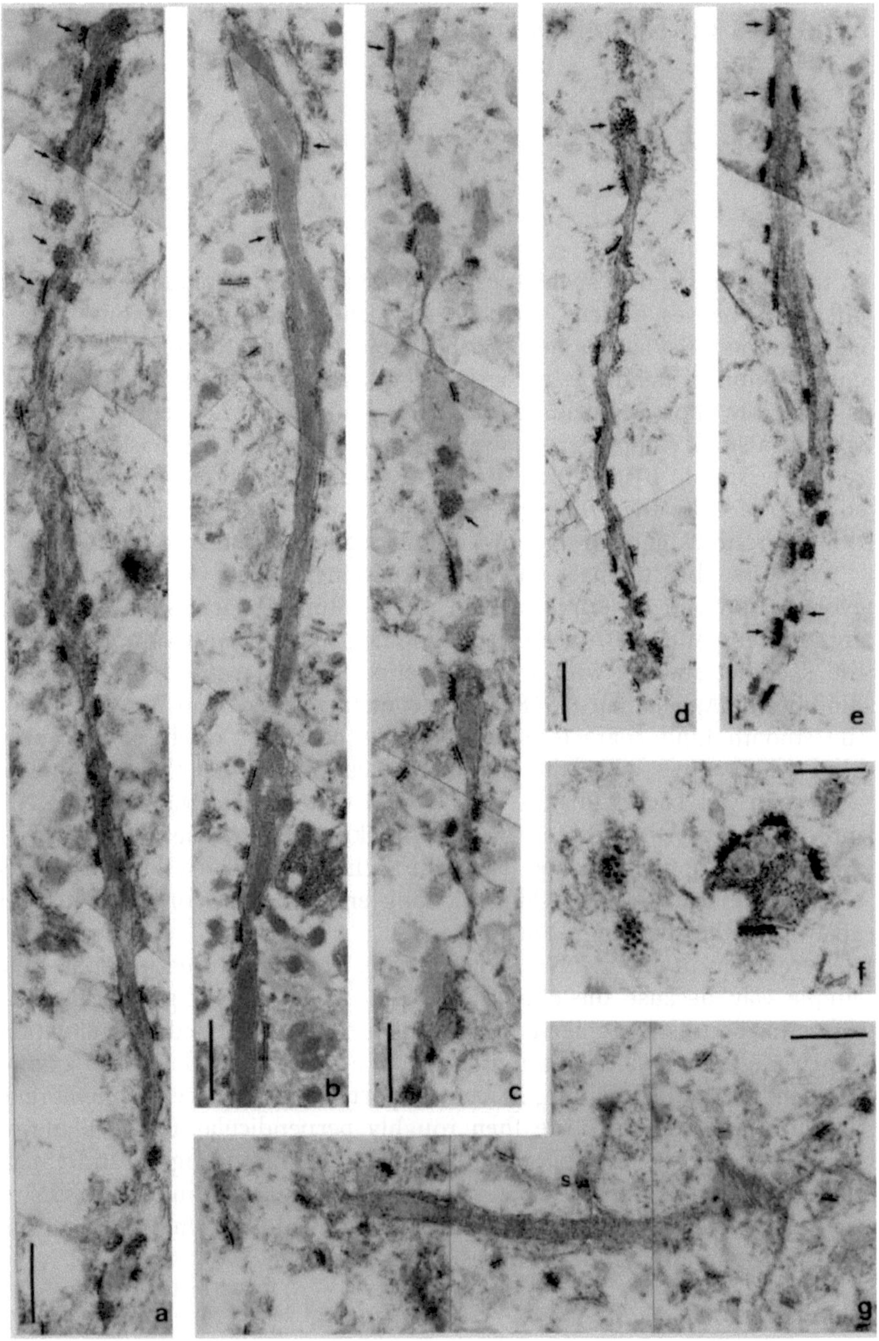

Fig. 44

The results, obtained from 24 pieces of dendrites taken from three mice with a total length of 201 μm, gave an average of 1.86 synapses per μm^2 of dendritic membrane, corresponding to *3.3 synapses per μm* of dendritic length (average thickness of the dendrites 0.56 μm). The average size of 91 synaptic junctions was 0.36 μm.

The number of synapses per length of smooth dendrite, 3.3/μm, compared to that of spiny dendrites, between 2/μm and 3/μm, is definitely in contrast with the idea of dendritic spines augmenting the number of synaptic contacts. The shrinkage in the case of the phosphotungstic acid preparations was checked and was found not to be responsible for the effect, since it was about the same as in the Golgi-celloidin preparations.

We must look for other explanations for the presence of spines on some of the cortical dendrites. There is one obvious difference in the distribution of synapses around spiny and smooth dendrites. In the latter case the position of the synapses is determined by the location of the dendrite, and this must influence the overall distribution of the synapses in the tissue. It is as if the smooth dendrites attracted the presynaptic sites towards themselves, and on the charts obtained from low power electron micrographs of our phosphotungstic acid sections (Fig. 46) one gains the impression that the regions around these dendrites (outlined by rows of synapses) are indeed impoverished of synapses. (We did not attempt to prove this quantitatively). On the other hand, the spines of the pyramidal cell dendrite may be thought of as devices which reach out into the tissue to make contact with unmovable presynaptic sites (Fig. 44 g). Recently we had reason to doubt this, as will be shown later (Schweizer, 1990). But it remains true that because of the spines the localization of synapses on the dendrites of pyramidal cells is less dependent on the geometry of the dendritic tree, forming as it were a fuzzy halo around it. This may be important in the statistics of the connections between axons and dendrites (Chap. 20). The probability of multiple contacts between one axon and one dendritic tree is certainly smaller in the case of a fuzzy distribution of the synapses around the dendrites, the spines of one neuron abundantly interlocking with those of many other neurons.

Fig. 44. Electron micrographs of sections stained with phosphotungstic acid. Some of the dendrites appear dark due to deliberate maltreatment of the tissue (ultrasound, suboptimal fixation). The dendrites marked *a–e* belong to spineless stellate cells. The section goes through the core of the dendrite in the upper part of dendrite *e* and grazes its surface on the lower part, showing the synapses in side view *(above)* and from the surface *(below)*. The dendrite marked g belongs to a spiny (pyramidal) cell. *f* is a cross-section through a spineless dendrite. Note that the smooth dendrite imposes its own order on the distribution of synapses in the tissue (see Fig. 46) while the spiny dendrite does not. Some of the synapses are marked by arrows. *s*: synapse on a spine. Bars: 1 μm (Schüz and Dortenmann, 1987)

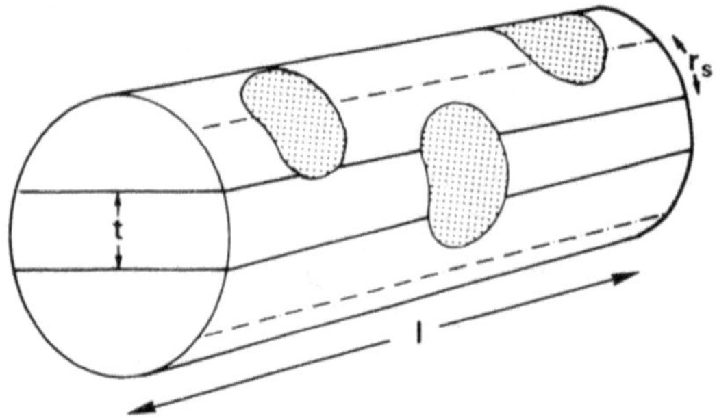

Fig. 45.
The geometry underlying counts of synapses on smooth dendrites. *t* is the thickness of the section, *l* is the length of the segment of dendrite considered, r_s is the radius of the synaptic thickenings stylized as roughly circular pancakes wrapped around the dendrite. (Schüz and Dortenmann, 1987)

Fig. 46.
Each *dot* represents a synapse on a section stained with phosphotungstic acid (according to Bloom and Aghajanian, 1968). On sections containing long pieces of smooth dendrites the position of the dendrite can be made out because of the alignment of the synapses. **a, b, c, d,** belong to four different sections. In **e** a region is shown for comparison, in which no smooth dendrite is contained in the section. Bars: 1 µm. (Schüz and Dortenmann, 1987)

It is easy to calculate, given the overall density of synapses and of spines, that within the sleeve that surrounds the spines of one given dendrite (Fig. 47) there must be almost ten times more spines belonging to dendrites of other neurons. A fibre entering that cylindrical region has a good chance (of about 9:1) of missing the dendrite in question, even if it does establish a synapse there. Note that by this reasoning the probabilities which we have calculated in Chapter 20 on the basis of Fig. 40, the "cross section" of the dendritic tree, turn out even smaller. Note also that

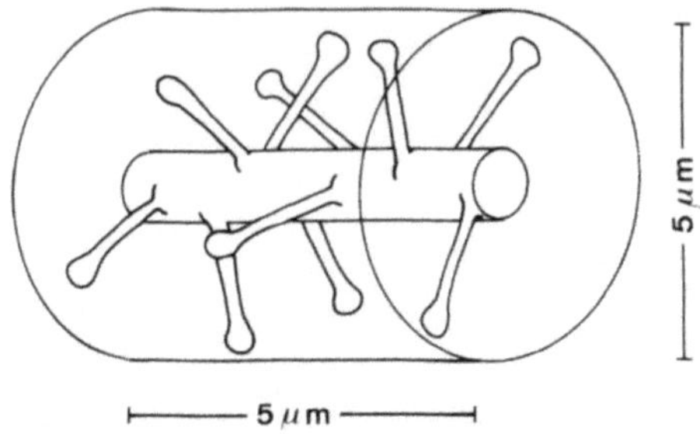

Fig. 47.
Distribution of spines around a spiny dendrite. The cylindrical region in which the spines are contained contains in addition about ten times more spine heads belonging to other neurons

with an average distance of 5 μm between successive synapses along an axon (Chaps. 8 and 9), an axon which establishes a synapse within the domain of one dendrite is likely to establish the next one outside, since the cylinder containing the spines of one dendrite measures about 5 μm across.

Thus spines do seem to contribute to the principle of maximum divergence/convergence which we have seen to hold in the cortex (Chap. 20), and it has even been suggested that this is their main purpose (Swindale, 1981).

However, this cannot be the whole story. We favour another explanation of the preponderance of spine synapses in the cortex, one connected with the subject of plasticity which we shall introduce in the following chapter.

Chapter 24

In Search of Engrams

Sensory input is projected onto the cortex in a fairly schematic way, well known to visual, auditory etc. physiologists. The motor output emanates from the cortex and reaches the effectors by a route which is again quite direct. We know that the effect of the input onto the output changes in the course of a life time, the change being called learning, conditioning and the like. This implies necessarily that the routing of signals through the cerebral cortex is variable, and as long as we consider fibres and synapses as the channels through which signals travel, we must conclude that learning (conditioning etc.) involves structural changes in fibres and/or synapses.

Although compelling, the idea of structural changes related to ongoing learning has not yet found unequivocal experimental support. Evidence has been produced for engrams (= structural changes due to learning or memory) at the level of molecules, at that of the macroscopical network of axons and dendrites or at any level in-between. None of this is compelling. Most experiments designed to produce detectable engrams in the brain used stimuli which are far too violent to be comparable to the information which normally circulates through the brain. "Sensory deprivation", e.g. complete deprivation of any visual input from birth, does seem to produce detectable effects (Valverde, 1967; see also review by Globus, 1975) but they may be indirect, perhaps mediated by a hormonal unbalance due to the artificial environment. Moreover they are sometimes more of the nature of a retardation rather than of a suppression of normal growth (Valverde, 1971; Winkelmann et al., 1977).

This does not imply that the structural changes involved in learning are so subtle that they escape light microscopic and even electron microscopic observation. On the contrary, it is well possible that with the conventional methods of histology we see engrams all the time, but are unable to relate them to the patterns of activity which produced them. The growth of an axon collateral or of a dendritic branch, the establishment of a synapse between two neurons or the increase in size of a synapse, the wrapping of a piece of axon in a myelin sheath, the increase of nucleic material or of protein in the cytoplasm are as many candidates for the elementary event which underlies memory. In addition, the disappearance or decrease of any of these cell constituents may also be used to construct models of memory processes. (For reviews see Greenough and Chang, 1988; Bailey and Kandel, 1993).

A major difficulty in the search for the anatomical traces of memory resides in the problem of separating the effects of growth from those of learning. Many animals, including those species which are most frequently used for experiments on memory, rats and mice, are born with very immature brains (Fig. 25). The growth of axons and dendrites within the grey substance and the establishment of synapses there happens to a large extent at a time when the sense organs are already exposed to the extrauterine world and nothing prevents us in principle from thinking that sensory information guides the growth processes. For the separation of genetically programmed development from learning it would be advantageous if at the moment of birth one was over while the other has not yet begun. Although the separation of the phase of development from that of learning is never sharp, there are animals, the so-called precocious ones (the others are called altricial) which largely meet this requirement, and among them one, the guinea pig, which is sufficiently close to mice and rats to warrant fruitful comparisons (Schüz, 1978, 1981 a, b, 1986).

The appearance of synapses and spines in the cortex has sometimes been interpreted as concrete evidence for the incorporation of memory into the brain. Figure 48 shows the increase of the number of spines per dendritic length as a function of age for the three species mouse, rat and

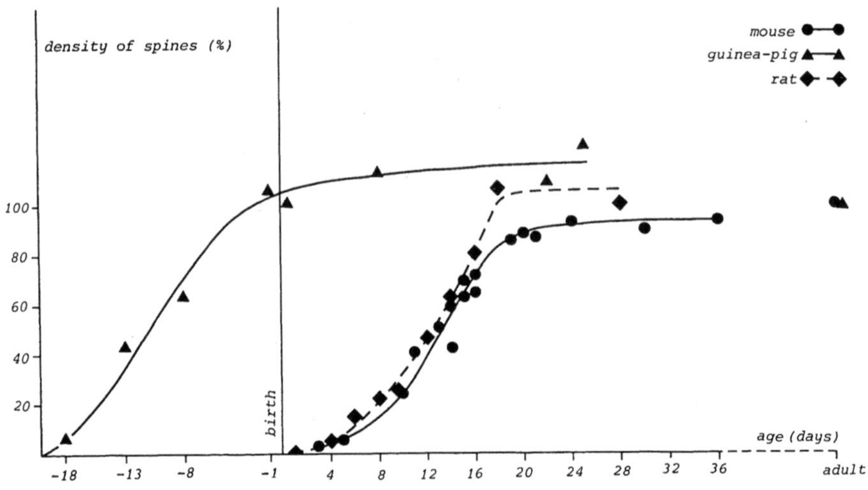

Fig. 48. Density of spines in the cortex of mouse, rat and guinea pig, as a function of age. The plot is in terms of percentage of the adult value. In the three species the spines are formed within 20 days with a very similar time course. Where the curve is steepest, about 50 000 spines are formed every second. All of this happens in the guinea pig during the 3 weeks preceding birth, in the rat and in the mouse during the 3 weeks starting with birth. It is unlikely, therefore, that the establishment of a dendritic spine reflects a learning process. Data for the mouse from Valverde (1971), for the rat from Schapiro et al. (1973). (Schüz, 1981a)

guinea pig. In the three species the time course is quite similar, the transition from almost no spines to almost their full number occupying about 20 days. But while this process begins at birth in mice and rats, it is almost completed then in guinea pigs. It might be more illuminating to redraw the three curves with the maximum slopes aligned: the statement could then be rephrased to the effect that mice and rats are born early and guinea pigs late on the developmental calendar.

It is interesting to note that the *rate* at which spines are produced, judging from the steepest slope of the curves in Fig. 48, is *between ten and hundred thousand per second*, far too many for the few bits of sensory input to be possibly of any significance in their placement. But more significant still is the fact that in the case of the guinea pig the great majority of spines (and synapses) are formed while the animal is in the womb, an environment which hardly offers much useful information to the growing embryo.

The increase in the number of synapses closely parallels that of the spines (Fig. 49), so much so that the comparison of the two curves provides no clue as to which comes first. In the case of the synapses residing on spines, is the synapse established first on the surface of the dendrite and then drawn or pushed out so as to end up on top of a spine, or do the spines reach out from the dendrites in order to meet their presyn-

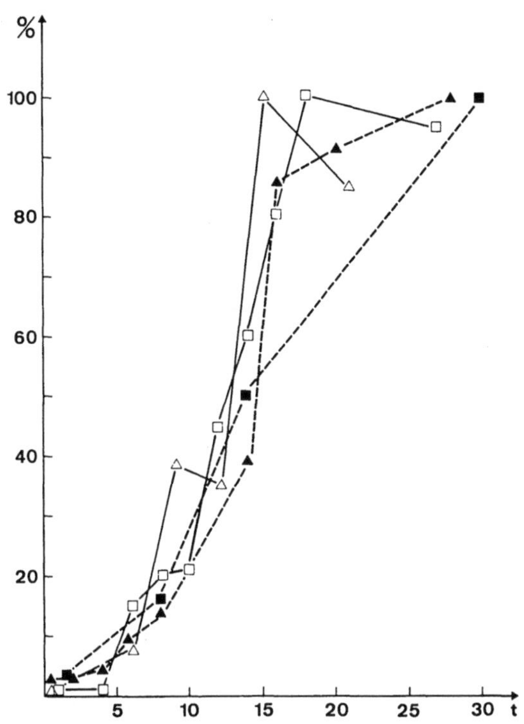

Fig. 49.
Postnatal increase of the number of spines *(white triangles and squares)* and synapses *(black triangles and squares)* in the visual cortex of the rat, in percent of the maximum value. The abscissa indicates postnatal age in days. (Data from Shapiro et al. 1973, Wolff 1978, Miller 1981, and Blue and Parnavelas 1983). The time course of the development of spines and of synapses is nearly identical, so that the question of which comes first cannot be decided by comparison of these data. (Schweizer, 1990)

a

b

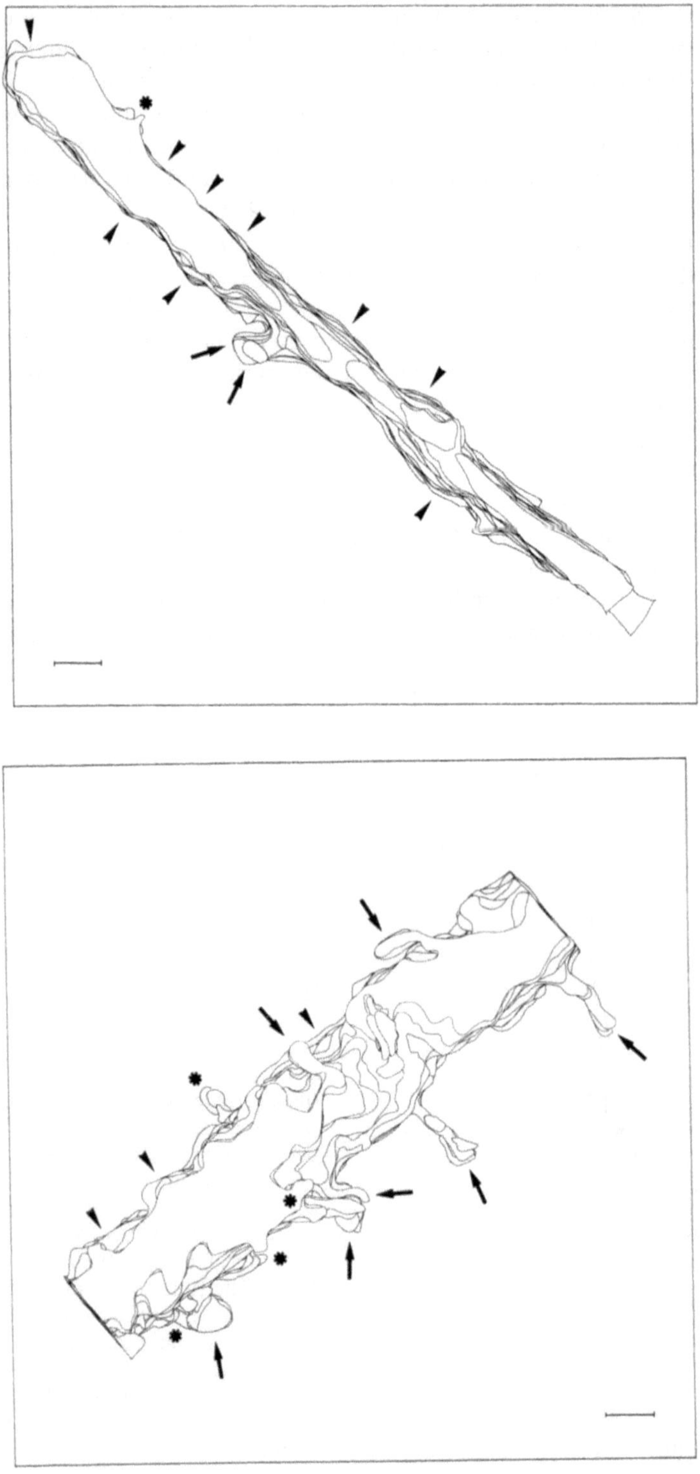

Fig. 50 a, b

aptic partners in the nearby neuropil and then establish a synapse there? The question was tackled experimentally by Schweizer (1990). Several dendrites in the cortex of a 12-day-old rat were examined in serial electron microscopic sections and compared to similar dendrites in a rat 21 days old (Fig. 50). There were very few spines on the 12-day-old dendrites, and more synapses residing on the dendritic shafts. In the older dendrites the spines had reached about their adult density and the number of synapses on the dendritic shafts were reduced in number, suggesting that some of the ones previously present had developed into spines. Moreover, not a single instance was found of a spine not carrying a synapse, as a further proof of the fact that when a spine grows it takes its start from a preexisting shaft synapse and carries it along with it.

In summary, the establishment of a synapse or of a spine in the cortical network is not very likely to be the elementary process involved in learning. Rather, the case of the guinea pig, where almost the full set of synapses is ready at birth, argues strongly for the idea that the presence of synapses is a necessary condition for the beginning of the learning phase. Mice and rats are no argument against it, since in the first week of extrauterine life, while their synapses are being prepared, their behaviour is hardly more than embryonic and their sense organs (at least their eyes) are not yet functioning. In these animals too, the completion of the synaptic setup inaugurates the active interplay with the environment.

Interestingly, this is not limited to the cerebral cortex. The cerebellar cortex is sometimes considered as being retarded with respect to the cerebral cortex, since the immature histology (the presence of an external granular layer) can be seen with the naked eye there. And yet the network in the molecular layer of the cerebellum is established quite in synchrony with that in the cerebral cortex, and with a similar time course both in mice and guinea pigs, in guinea pigs before birth, in mice after (Schüz and Hein, 1984).

The search for anatomical traces of memory must continue at a different level.

Fig. 50 a, b. Reconstruction of dendrites of pyramidal cells from serial electron micrographs. **a** rat, 12th postnatal day; **b** 21st day. During this period spines develop. It can be seen that in the younger dendrite *above* there are more synapses on the dendritic shaft *(arrowheads)* and only two on a dendritic spine *(arrows)*. In the older dendrite *below*, there are only three synapses on the dendritic shaft and eight synapses on spines. It is very likely therefore that synapses are first formed on the shaft and then the spines protrude carrying the synapses with them. *Asterisks* indicate truncated spines. Bar: 1 μm. (Schweizer, 1990)

Chapter 25

Postnatal Changes, Possibly Due to Learning, in the Guinea Pig Cortex

The guinea pig is born with the adult colouring, with functioning sense organs, with surprisingly good motor coordination and vivacious behavior setting in only minutes after birth. The brain is about 35% smaller than in the adult, but its internal structure looks the same down to the level of the network of fibres and synapses. And yet, the guinea pig learns as readily as related species do (Jonson et al., 1975; Petersen et al., 1977) and we expect therefore structural changes to take place in its brain after birth perhaps to the end of the animal's life. Such changes accompanying learning, neatly separated from the effects of development by the event of birth, should be more readily detectable in the guinea pig than in other species.

With this idea in mind, the study of the prenatal development of the guinea pig's cortex (Schüz, 1981a; see Chap. 24 here) was followed up with a study of postnatal changes (Schüz, 1981b, 1986). Whatever is added to the anatomy of a newborn guinea pig brain in the course of its life, or subtracted from it, may be taken as a plausible substrate of memory.

A closer look at the number of spines in the cortex, which to a first approximation we have seen to reach its adult value at birth, may still hold surprises for us. True, there is not enough flow of information in the sensory channels to determine the establishment of 10^4 or 10^5 synapses in a second, but the information may suffice to delete some of them selectively in the process of learning. Such learning by an initial excess and subsequent reduction of the neuronal connnections has been postulated and evidence in support of the idea was found in the human cortex (Huttenlocher, et al. 1982), in the olfactory bulb of the ferret (Apfelbach and Weiler, 1985; Apfelbach, 1986) and in the acoustic system of the chicken (Wallhäuser and Scheich, 1987). With respect to the guinea pig, a look at Fig. 48 also hints at a lower density in the adult than in the newborn. However, the effect is so small that it might simply be due to postnatal growth of dendrites. Fig. 51 is an attempt to estimate the relationship between dendritic growth and total number of spines after birth. The values in this figure are based on the densities in Fig. 48 and on the assumption that the dendrites of pyramidal cells grow postnatally as much as the cortex does in one dimension (which is 13% in thickness and about 18% in length and width). The curve shows a slow increase in the total number of spines during the first weeks after birth followed by

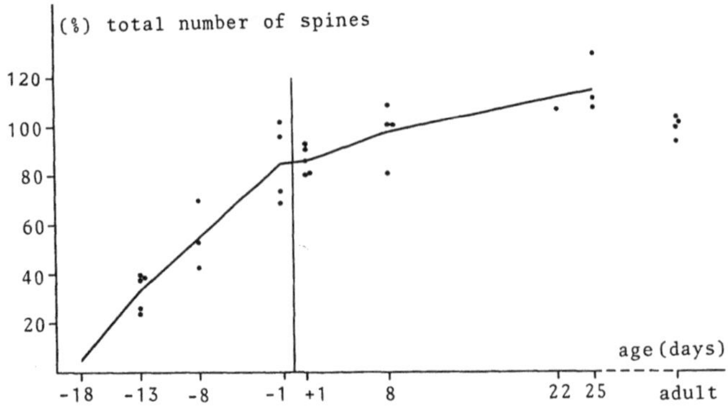

Fig. 51. Estimate of the *total number* of dendritic spines (in percent of the average adult value) in the cortex of the guinea pig as a function of age. The values are based on the densities per dendritic length in Fig. 48, multiplied by a factor for the estimated increase in dendritic length. Each dot corresponds to one animal (Schüz, 1981 a)

a decrease toward the adult figure. Both effects are not spectacular and we have no way of showing or denying their significance in memory. Neither can we confirm or exclude the possibility of a turnover of spines or synapses in adult life (Wolff et al., 1995).

If the number of spines and synapses provides no reliable clue to the mechanism of memory, the fine structure of these elements may. The first guess is the size of the synapses, intended as the surface of those membrane specializations on both the pre- and the postsynaptic side which are readily recognized in the electron microscope (Figs. 13, 42). The length of the postsynaptic thickening was measured on sections of guinea pig cortex of animals 1 day before (presumptive) birth and of adult animals. Although this length reflects the size of the synapse only statistically, depending as it does also on the geometrical relation between the section and the synapse, a difference in size in the two samples should appear from the statistics of the measured lengths. The difference was not significant in samples taken from posterior regions of the cortex, while in the anterior samples the adult synapses were slightly larger than the perinatal ones. Fig. 52 shows the overall-result (the data from anterior and posterior region are lumped together in Figs. 52, 53, 58, 59).

The difference between adult and prenatal synapses was more striking when the vesicles were considered on the presynaptic side (Fig. 53). The older synapses contain definitely more vesicles than the unborn ones, the largest numbers of vesicles per synapse being found practically only in the adult case. What may be even more significant is the increase in the variance of the number of vesicles (Fig. 54). This is to be expected in a sample of elements, some of which have undergone some change because they have served as memory traces, while others have remained un-

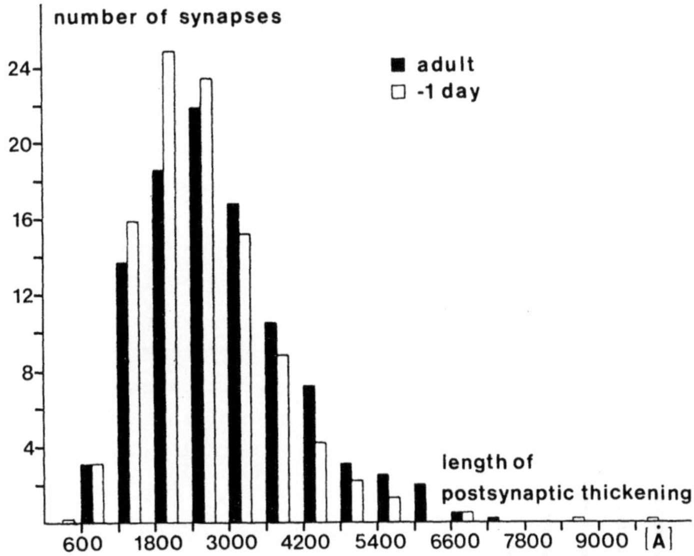

Fig. 52. Comparison of the frequency distribution of the different lengths of postsynaptic thickenings in adult *(black columns)* and just prenatal guinea pigs *(white columns)*. The two distributions are not significantly different. (Schüz, 1981b)

changed. If between birth and adulthood we observed some changes which are quite the same in all the elements, we would refer these to growth, or ageing, rather than to learning. If information is to be stored, differences must necessarily be introduced among the individual carriers of this information.

The other impressive difference between newborn and adult guinea pig cortex concerned the shape of the dendritic spines. This is evident even to the inexperienced observer, once his attention has been drawn to the fact that in young brains the spines tend to be more slender and delicate than in the older ones (Fig. 55). Again, as in the case of the number of synaptic vesicles, what makes the shape of spines a likely candidate for the anatomical substrate of memory is the variety within the population of spines: in the adult dendrites some of the spines have grown quite chubby, while others preserve their juvenile appearance. The possible importance of this finding prompted a more quantitative study (Schüz, 1986), the results of which confirmed the intuitive observations.

There are considerable technical problems inherent in the small size of the spines. The geometry of individual spines can be assessed with precision by electron microscopy on serial sections, but this method is so time-consuming that its application to a sample sufficiently large for statistical evaluation becomes prohibitive. Thus high-power light microscopy on Golgi preparations has to be used, but the resolution of this method has its physical limits in the wave length of visible light (around 0.5 μm),

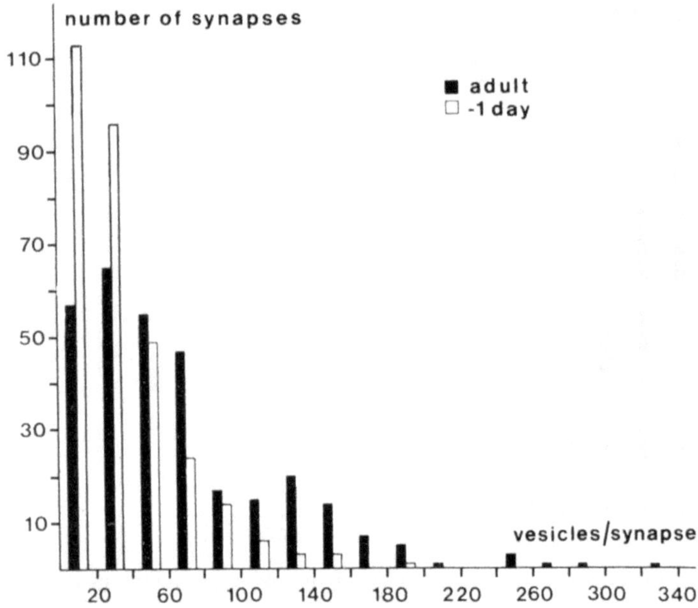

Fig. 53. Frequency distribution of the number of vesicles per synapse in adult *(black columns)* and just prenatal guinea pigs *(white columns)*. There is a significant increase in the number of vesicles with age. Note the bimodal distribution in the adult histogram which may indicate two kinds of synapses, perhaps such that have learned and such that have not yet learned. (Schüz, 1981b)

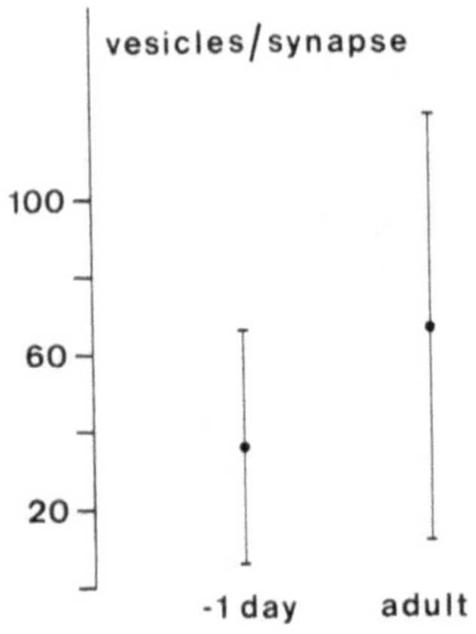

Fig. 54.
Average and standard deviation of the number of vesicles per synapse in prenatal and adult guinea pigs. Note the increase in the standard deviation. (Schüz, 1981b)

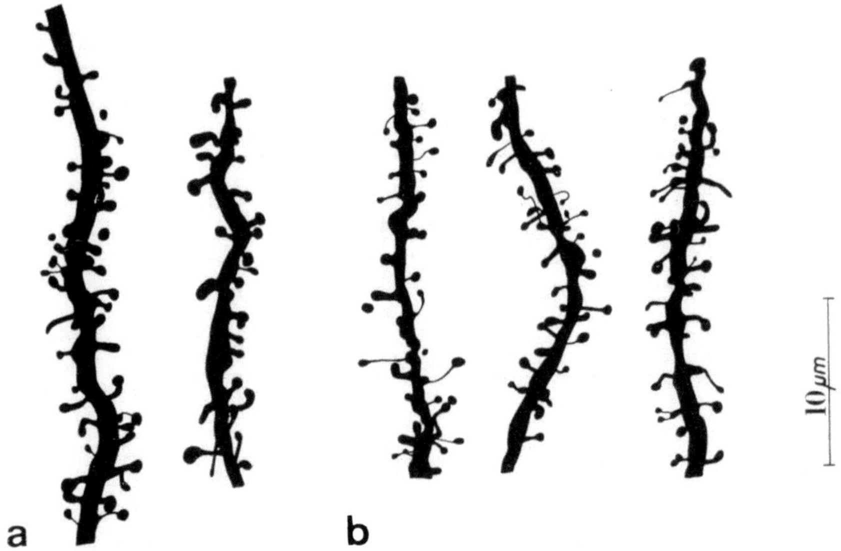

Fig. 55 a, b. Fragments of spiny dendrites of adult guinea pigs (**a**) and guinea pigs around birth (**b**). Camera lucida tracings of Golgi pictures. The thicker spines are more frequent in the adult sample. (Schüz, 1981b)

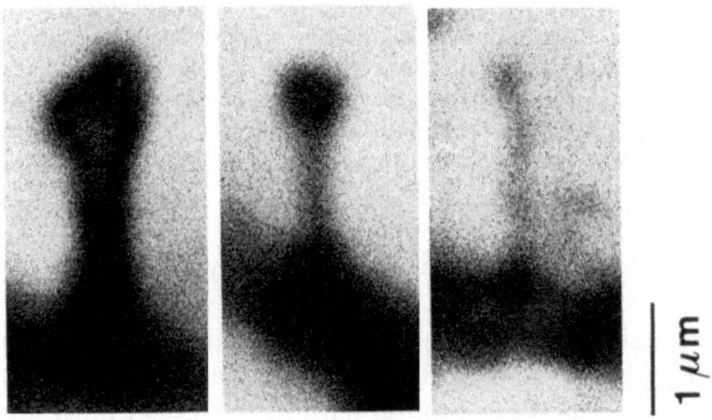

Fig. 56. Microphotographs of dendritic spines of different calibre. The enlargement is greater than that which the resolution of the optical system would warrant. Still, the intensity of the shadow makes it possible to evaluate the thickness of the spine. (Schüz, 1986)

while some of the slimmer spines have necks with diameters not exceeding 0.1 µm (Fig. 56). However, the thinner spines are not invisible, only their outlines are ill-defined. If we suppose that they are all uniformly filled with silver chromate which precipitates in the Golgi stain, we can get a fair indication of their size by the intensity of the shadow which appears on the microphotographs. In practice, on high contrast prints, the core of the spine may appear black even in the thinnest spines and its width can then be taken as an indication of the real thickness of the spine (Fig. 57). Since the aim was only a relative evaluation of the sizes of

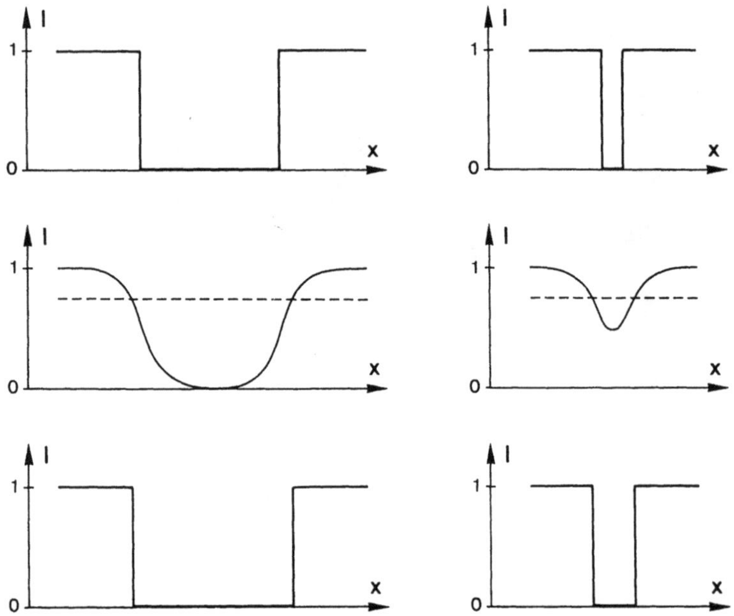

Fig. 57. To show how the size of a small object may be represented on a photograph even if the resolution of the optical system is insufficient. *Upper row* the distribution of the luminosity *(I)* of an object along a spatial coordinate *x*. *1* represents white and *0* black. The object on the *left* is larger than the wave length of the light, the one on the *right* is smaller. *Middle row* distribution of luminosity on the focal plane of the optical system. The shadow cast by the smaller object on the right is nowhere black. *Lower row* a high contrast photograph, with a cut-off level (i. e. discrimination between black and white) as indicated by the *dashed line in the middle row*, restitutes the full contrast to the image, whereby the size of the object is roughly represented

spines, the error inherent in this method does not really affect the basic result.

The heads of the spines are indeed bigger in the adult sample (Fig. 58), the difference being statistically significant at the level of $p < 0.001$. The most frequent values are around 0.6 μm in the newborn and around 0.8 μm in the adult cortex. Similarly, the thickness of the spine stalks is increased in the adult animals (Fig. 59) with a statistical significance of $p<0.01$. On the other hand, the impression that the spines are longer in the young cortex is not convalidated, and is probably due to their more slender shape. There is no detectable correlation between the size of the spine head and the thickness of the stalk either in the newborn or in the adult sample, and no correlation between the length of the spine and the other measurements.

These measurements on spine dimensions agree fairly well with similar measurements on Golgi preparations by Jacobson (1967). Other studies using the electron microscope have also reported comparable values for

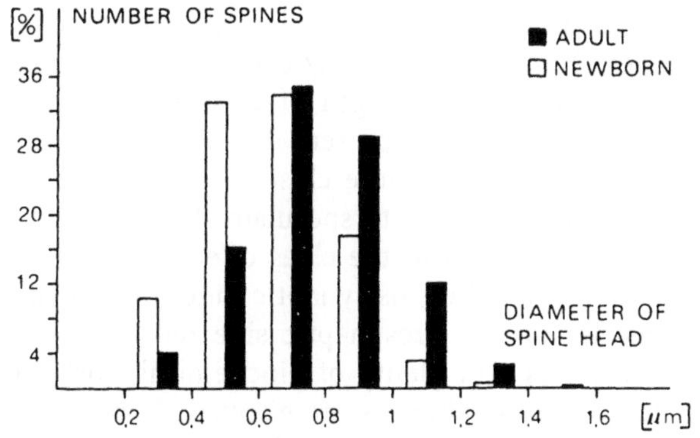

Fig. 58.
Frequency distribution of the size of the spine heads in adult *(black columns)* and newborn *(white columns)* guinea pigs. The thicker heads are much more common in the adult. (Schüz, 1981b)

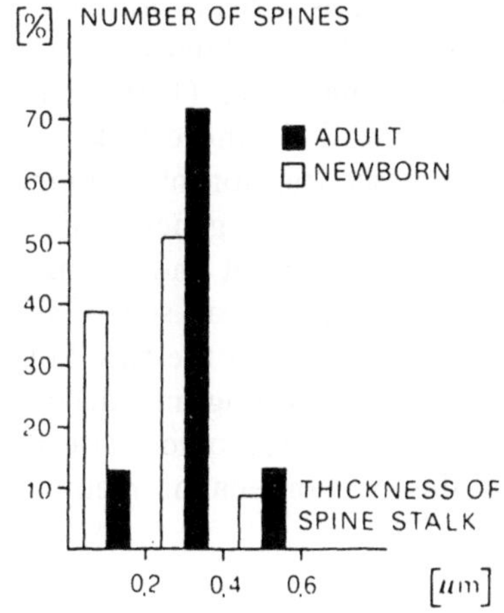

Fig. 59.
Frequency distribution of the thickness of spine necks in adult *(black columns)* and newborn guinea pigs *(white columns)*. The thinnest necks are more frequent in the newborn sample. (Schüz, 1981b)

the thickness of spines in the hippocampus (Fifková and Anderson, 1981), and some slightly higher values in the striatum (Wilson et al., 1983), the difference being probably due to the technical difficulties.

We conclude our search for the *anatomical traces of learning* with *two likely candidates*, the *number of vesicles* on the presynaptic side of a synapse, and, for the synapses involving spines, the *thickness of the spine*. Both have been proposed earlier, the number of vesicles by Garey and Pettigrew (1974), Vrensen and de Groot (1974), and the shape of the spines by Globus and Scheibel (1967b), Fifková and Anderson (1981). (For reviews on spines and plasticity, see Calverley and Jones, 1990; Horner, 1993). The interpretation of spines as memory devices has prompted F. Crick to propose a theory of memory in which spines play a dynamic role, and biochemistry compatible with the idea of active deformations of the spine by some kind of contracting mechanism has been reported (Fifková and Delay, 1982).

Obviously there are molecular mechanisms underlying the phenomena which are observed in histology. Nor can we entirely exclude that the essential molecular device which makes the strength of a synapse dependent on the history of its past activity escapes present-day histology, even if the electron microscope affords a picture quite close to the molecular dimensions. But it seems somewhat gratuitous to speculate about molecular mechanisms of memory without regard to the clues offered by histology. If the thickening of the spine on the postsynaptic side, or the increased size of the bag of vesicles on the presynaptic side reflects some elementary learning process, there is still plenty of biochemical work to be done in order to elucidate how these changes are effected. In particular, if "Hebbian learning" is what happens there, we should like to know how the combined pre- and postsynaptic activity translates into molecular changes. A great deal of biochemical evidence is already available, both in vertebrates and invertebrates: Byrne (1985), Collingridge and Bliss (1987), Wickens (1988), Kennedy (1989), Malenka et al. (1989), Bailey and Kandel (1993), Griesinger (1997). Such work is more welcome than the repeated attempts at salvaging an old misconception of memory being inscribed into the structure of macro-molecules (like genetic information in DNA). Since we know that both the sensory input and the motor response are coded in the cortex in the form of widespread sets of active neurons, how could this information be compressed in the tiny bottleneck of the structure of individual molecules? The coding and decoding processes between molecular structure and macroscopic patterns would seem to be more complicated than the phenomena of memory which they are supposed to explain.

Chapter 26

Cortico-Cortical Connections

Up to this point, our essay has concentrated on the connections between cortical neurons within a neighbourhood, say, 1 or 2 mm across, corresponding to the length of the longest horizontal axon collaterals plus the length of the longest dendrites. The influence which various kinds of stellate cells exert on the cortical network is indeed limited to a range even smaller than that, but many pyramidal cells have axons leaving the cortex for the subcortical white substance and re-entering the cortex in distant places. We estimate that this long-distance system of connections is about as powerful as that of the local connections, the estimate being based on a comparison between the apical and the basal dendrites of pyramidal cells. Judging from the relative size of the two portions of the dendritic tree, both receive about the same number of synapses, the first, we may assume, prevalently through long-range connections and the second through local ones. In a rough way this must be so, since the fibres mediating the long-range connections tend to terminate in the upper layers of the cortex, where apical dendrites from all layers ramify, while the local axon collaterals are distributed mostly below or at the level of the cell body and thus enter a neuropil consisting largely of basal dendrites of neighbouring cells.

It is not clear whether long-range and short-range connections between pyramidal cells play entirely different roles or whether they are just the extremes of a continuous distribution, some fibres making their way through the white substance perhaps only for convenience of construction during development or to minimize distances, while the shorter ones for similar reasons travel toward their target neurons directly through the cortical grey substance. It is also not clear just what degree of predetermined order governs the system of cortico-cortical connections in the white substance. One thing is certain, there are well-defined pathways, some of them forming macroscopical bundles in larger brains, which connect cortical areas to each other and definitely contradict the naive view of a complete or random set of connections between all points of the cortex.

Basically, the white substance of the telencephalic hemispheres is made up of the following sorts of fibres: (1) afferents from, (2) efferents to subcortical stations, (3) commissural fibres and (4) fibres connecting different points of the same hemisphere. The external connections of the cortex (1 and 2) were not our concern in the present study, but the literature over the years has produced many reports, previously based on degeneration techniques and lately on the injection and active transport of various substances. The commissures (3), especially the fibres of the corpus callosum, were a favorite subject of some studies on long-range connections, in particular on rodents: Hartenstein and Innocenti (1981), Hedreen (1981), Jacobson and Trojanowski (1974), Yorke and Caviness (1975). The so-called associational fibres of the white substance (4) were studied mainly on larger animals with injection techniques (review in Pandya and Yeterian, 1985), but some studies on rats were also published (Montero et al., 1973; Isseroff et al., 1984; Miller and Vogt, 1984), primarily involving the visual areas. Most of these studies were aimed at the definition of orderly projections of one cortical area onto another, and such orderly mapping was indeed found in most cases. In addition, specific connections were sometimes found that involved regions smaller than the cytoarchitectonic areas (Miller and Vogt, 1984), and in general the projections tended to be patchy rather than continuous.

A study by Greilich (1984) in our laboratory was designed to reveal a possible global pattern in the system of cortico-cortical connections of the mouse. Quite apart from the patchy appearance of some of the projections, it was still an open question whether the overall picture was one of widespread connections of every part of the cortex with every other part, or some more restricted scheme. Greilich examined 21 brains of mice, each of which had received an injection of horseradish peroxidase (HRP) restricted to about 1 mm^2 of cortex. The injection sites varied and were chosen so as to let the entire sample cover as completely as possible the whole dorsal surface of the cortex (roughly corresponding to the "neocortex", or to the region with numbered areas on our map in Fig. 74). HRP travels from the axon toward the cell body, so that after an appropriate lapse of time it fills (and allows to stain) the cells whose axons reach the injection site.

In every instance most of the cells projecting to the injection site were localized in discrete patches, the size of which was roughly equal, or smaller than that of the region injected. Nearly always there was more than one patch, in some cases up to eight separate patches widely distributed over the cortex. There was an unexpected finding: with most injection sites there were not any stained cells in the cortex below the rhinal sulcus (see Chaps. 2 and 31). This indicates that the part of the hemisphere roughly corresponding to what is sometimes called the allocortex, does not participate on an equal footing in the intimate interplay of the neocortical areas. This is confirmed, rather than contradicted, by recent evidence of a diffuse projection from the entorhinal area to the entire cortex obtained with anterograde transport methods (Swanson and Köhler, 1986).

Within the neocortex (= the dorsal part of the hemisphere) there were regions which received fibres from many places and other regions which were recipients of only few projections. Figure 60 displays this for 12 dif-

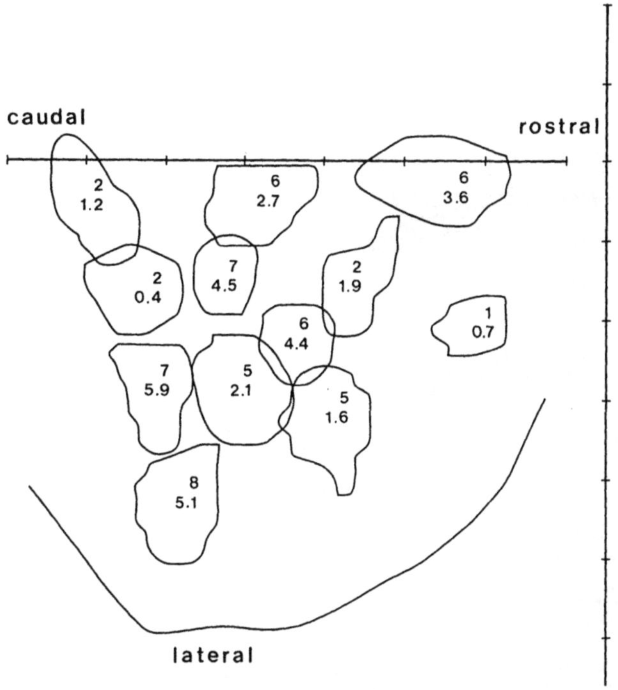

Fig. 60. Convergence of cortico-cortical projections. The chart represents a planar map of the dorsal and medial hemisphere of the mouse (caudal to the left, lateral down; the *horizontal line* represents the dorsal rim of the interhemispheric sulcus, where the dorsal and the medial cortex meet). The *closed lines* show the sites of 12 different injections of horseradish peroxidase. The *upper number within each site* indicates the number of separate patches in which the cells of origin projecting to that site were localized. The *lower number* indicates the size of the total area in which these cells were localized, expressed as the ratio of that area to the area of the corresponding injection site. Thus, for example, the numbers in the *upper right corner* indicate that cells of origin were found in six separate patches, and the total area of these patches was 3.6 times larger than the area of that injection site. (Greilich, 1984)

ferent regions, indicating the number of patches in which the cells were housed which projected to any particular region. We may call this the variation of the convergence over the surface of the cortex. Conversely, divergence can also be defined as the number of regions to which fibres project from any one location in the cortex. Figure 61 displays this graphically, with the size of the black dots representing the degree of divergence. There are centres of maximal divergence which, peculiarly, correspond largely to the centres of maximal convergence. A region in the posterior hemisphere is prominent both on the divergence and on the convergence map. Its position makes it possible to identify it with the primary acoustic area, area 41 on the Caviness map (see Chaps. 30, 31, Figs. 66 and 74).

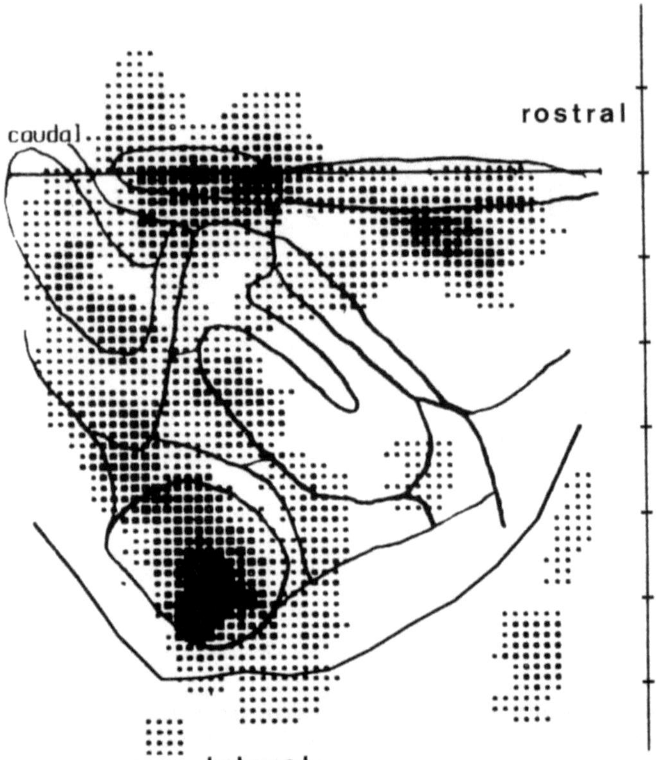

Fig. 61. Divergence of cortico-cortical projections. The chart is similar to that of Fig. 60 but represents the same data in an inverse way. The *size of the black dots* indicates the number of injection sites which are reached by cell bodies located in the corresponding region. The *largest dots* indicate divergence to six different areas. The outlines of cortical areas in the Caviness map are also shown. (Greilich, 1984)

Figure 62 displays the distribution of distances between the origin of a projection and its target region (i.e. between the cells stained and the injection site) in form of a histogram. The distance varies from less than 1 to 7 mm. The shape of the histogram is compatible with a fairly continuous distribution of distances, from those comparable to the range of intracortical axon collaterals (a few tenths of a millimeter) to the longest, which span almost the entire hemisphere. The argument that intracortical and cortico-cortical connections between pyramidal cells are not really different except for their length, is neither proved nor disproved by these statistics. But the overall picture is one that suggests a high degree of macroscopical order in the cortico-cortical connections, a higher degree of order, most likely, than that in the system of local connections which we have described in the previous chapters.

It is interesting to compare the reality as described by Greilich (1984) and by many others, with an abstract scheme we had once proposed

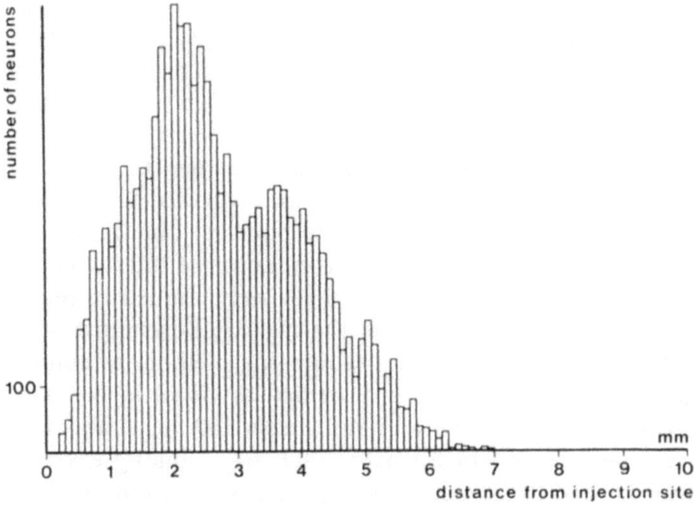

Fig. 62. Frequency distribution of the distances spanned by the cortico-cortical connections illustrated in Figs. 60 and 61. Only homolateral connections are shown. The longest cortico-cortical fibres found in this study are about 7 mm. There is an indication of a second peak in the histogram whose significance is obscure. (Greilich, 1984)

(Braitenberg, 1978b). The question was, how much white substance one would expect to find in brains of animals of different sizes, if a full set of cortico-cortical connections were established there. Since a direct connection of every neuron with every other neuron is obviously out of the question, we may look for a parcellation of the cortex such that one could assume every *compartment* to be connected to every other one. If each (pyramidal) cell produces exactly one cortico-cortical axon, a reasonable parcellation would be into $\sqrt{N}$ compartments, each containing $\sqrt{N}$ cells (N being the total number of cells). One could then assume every compartment to send an axon to each of the others. There would be about 3000 such "square root compartments", each containing 3000 neurons in the case of the mouse, and about 100000 compartments each containing 100000 neurons in the case of the human cortex. It is interesting that the size of the hypothetical compartments, i.e. the diameter of a cylinder of cortex containing 3000 or respectively 100000 neurons, is 0.17 in the mouse and 1 mm in the human, roughly the same as that of the largest dendritic trees in both species. Also, the human "square root compartments" correspond in size to various kinds of "columns" or "modules" which were described in primate (and cat) neurophysiology. Interestingly, the amount of white substance which one expects by assuming a full set of connections between $\sqrt{N}$ compartments corresponds rather well to the observed volume of the white substance in mouse, monkey and man. The point should not be stressed too much, however, since we have already seen that if there are any discrete compartments in the cortex, the connections between them are far from being homogeneous. Rather, the idea of $\sqrt{N}$ compartments can be taken as an abstract version of a suspicion which will come up again talking about cortical areas. It seems that the full set of connections within the cortex is achieved in two stages, macroscopic connections between areas (compartments), and connections within each area (compartment).

It is a fact, that, at least in larger mammals, the structure of the white substance is best described as a set of discrete bundles joining areas or parts of areas to each other, rather than as a continuum of intertwined fibres. This interareal traffic has been most exhaustively studied in the visual system of the macaque monkey and of the cat (e.g. Rosenquist, 1985; Casagrande and Kaas, 1994). There is a discouraging plethora of connections among the 30 or so cortical areas described as visual in the monkey, numbering about 300 (Felleman and Van Essen, 1991), i.e. about 1/3 of the number of connections which would result from joining every visual area with every other. Faced with this situation, in an attempt to discover the overall scheme, Young and his colleagues (Young, 1992; Scannell at al., 1995; Young et al., 1995) embarked on a heroic review of papers dealing with cortico-cortical connections in the entire hemisphere in both cat and monkey. They found 1139 connections between 65 areas in the cat cortex and 834 connections between 73 areas in the monkey cortex – 27% and 15%, respectively, of all possible connections (Young et al., 1994). The data were then transferred onto charts in which the "proximity" of cortical areas is rendered in an abstract space, where it is not the metric of anatomical space that counts but rather the functional vicinity, as evidenced by unidirectional or reciprocal fibre connections. One of the most surprising results of this effort is that the abstract space thus defined is not so very different from the actual space of the cortical layout, demonstrating that on the whole the dominating rule is "propinquity breeds matrimony", in other words, a connection between two areas is more likely to exist when they are geometrically close to each other. Or, alternatively, the functional relatedness embodied in fiber tracts leads to anatomical proximity by some principle of minimization of distances (Van Essen, 1997).

Chapter 27

Cortical Architectonics

In the first part of our essay (Chaps. 3 to 13), we described the cortex as a mixture of axons, dendrites and synapses of various sorts, without any consideration of the constraints which the shapes of the neurons impose on the statistics of their connections. We took care of these aspects later (Chaps. 14 to 21) when we considered the contribution of individual neurons to the cortical neuropil. Even so, the picture of the cortex we had in mind was that of a large continuous network of interconnected neurons, without any internal boundaries which would justify an interpretation in terms of separate subunits or modules. Only in the preceding chapter (26) did we produce some evidence for smaller subdivisions of the cortex, 1 mm or so across, which seem to preserve their separate identities at least in the pattern of long-range connections which they establish with each other. These are very likely related to the "cortical areas" of the older descriptions, each characterized by a peculiar, often only slight variation of the general cortical histology, and in many cases also by a special relation to input or output fibre systems. We have seen that area 41, known as the auditory field, is also characterized by massive cortico-cortical connections (Figs. 60, 61).

An investigation of how the statistics of neuronal connections vary in different areas would be the natural follow-up of our work up to now, and one which could both confirm and extend our conclusions. However, this is an endeavour which would require indefinite time with the present techniques. The Golgi work hardly provides valid statistics even if the results for the entire cortex are lumped, and has never contributed enough information about any particular area to serve as a basis for a convincing functional model, the visual areas of primates not excluded. Other methods, e.g. by systematic injection of dyes in many neurons, may hold such a promise but have not developed into routine procedures yet. Thus we must be satisfied with a qualitative look at the areal diversity such as the so-called cortical architectonics provides.

The term architectonics designates a trend in neuroanatomy which flourished around the turn of the century and in the following decades. The common denominator between the various schools of architectonics, in particular architectonics of the human cerebral cortex, was inspection of histological preparations at low magnification, by which means subtle local differences in the distribution of cell bodies, fibres and other tissue elements could be detected, as well as local variations in the statistics of neural shape and size. Since the techniques used were almost exclusively staining of cell bodies according to one of the Nissl methods and staining of myelinated fibres, a connection of these findings with the results of Golgi analysis was difficult to obtain and was frequently not even sought.

The episode of cortical architectonics ended in a stalemate situation after a fierce argument between two opposing parties, the one claiming a parcellation of the human cortex in a multitude of individual, very sharply demarcated "areas", the other stressing common traits and continuous transitions between larger regions of admittedly different fine structure. (O. Vogt and his school on one side, Bailey and v. Bonin on the other).

We are now ready to reconsider the problem of cortical areas in the light of two recent developments. First, the advances of cortical electrophysiology in sensory areas of various mammals brought to light a mosaic of sensory maps, and within many areas a mosaic of smaller subunits (columns and hypercolumns) which goes beyond the expectations of even the most fanatic upholders of architectonic parcellation. Secondly, we are beginning to observe plasticity of cortical structure under the influence of the environment, in other words, changes in the arrangement and shape of neuronal elements that can be correlated with various sorts of sensory input. It becomes imperative, then, to distinguish between the effect of such plasticity, possibly different in different places of the cortex, and inborn local variation of the cortical wiring. In a way, we may interpret areal variations, when they are genetically determined, as the inborn conditions for learning in various (visual, auditory, linguistic, motor, etc.) contexts, or if you wish, as predetermined knowledge about these contexts incorporated in the brain through the genetic channel.

For the following sketch of areal architectonics of the mouse cortex we proceeded as follows. We began by scrutinizing our own preparations as well as the SAP atlas without previous study of already existing cortical maps, in search of easily recognizable boundaries between regions of different structure. These were marked on our own map (Fig. 73). We then read the following papers which deal with areal architectonics in the mouse: Isenschmid, 1911; Rose, 1929; DeVries, 1912; Drooglever Fortuyn, 1914; Rose, 1919; Caviness, 1975; and for comparison some papers on the rat cortex: (e.g. Krieg, 1946, and Schober and Winkelmann, 1975). We found it difficult to establish an unequivocal picture of areal variation from the literature, since the reports by different authors are at variance not only as far as the interpretation of the areas (e.g. as acoustic, or somatosensory or visual) is concerned, but more distressingly so also about their delimitation, and to a lesser extent, about their description. However, where the descriptions are detailed enough, it was possible most of the time to establish concordance between the authors as well as with our own observations, which became much more detailed after these older descriptions had acted as an eye-opener. We shall now compound our own summary "architectonics" and briefly return to the comparison with the literature at the end.

Chapter 28

Layers

The existence of several distinct layers of the cortical grey substance, arranged in parallel to the surface of the cortex is obvious even to the naive observer (Figs. 63, 64, 66). Another question is how many of these layers there really are and another one again, what are their distinguishing characteristics and how can one and the same layer be identified in cortical areas having very different architectonic structure. Generally, authors agree in their distinction of six cortical layers, homologous even between different species such as mouse and man, and we have no good reason for not following the established terminology. It is important to realize, however, that on the basis of more refined knowledge which came with the introduction of new staining techniques, if we had to take a fresh start today, we would probably use a different numbering, witness the sublayers and sub-sublayers that had to be introduced in some cortical areas.

One of the difficulties that mars any attempt at numbering layers in the cortex is the fact that the majority of cortical elements is not confined to any one layer. Thus, obviously the fibres reaching the upper echelons of the cortex from the white substance, and those leaving the cortex from its upper layers, traverse all layers and so do the long apical dendrites of most pyramidal cells of the lower layers. Generally the dendritic trees of cortical neurons do not seem to respect the borders between layers, nor do many of the axonal ramifications. Still, the stratification that is apparent to the architectonic eye looking at the cortex at low magnification must correspond to something meaningful, and is in any case at the basis of the most important areal distinctions.

In order to convince ourselves of the reality of the cortical layers (and sublayers), we looked in the mouse cortex for regions in which the borderline between two adjoining layers was easily and unequivocally to be seen. Even if, in some other area, the same border was not so clear, it could be recognized there too by horizontal homology, i.e. by following through the layers horizontally from one area to the next. The following borders between adjacent cortical layers were well marked in at least one area:

The Border Between Layer I and II. This is the level which is most clearly defined everywhere in the cortex (e.g. Figs. 63, 64, 65, 66). There is an abrupt increase in the density of neural cell bodies as one proceeds

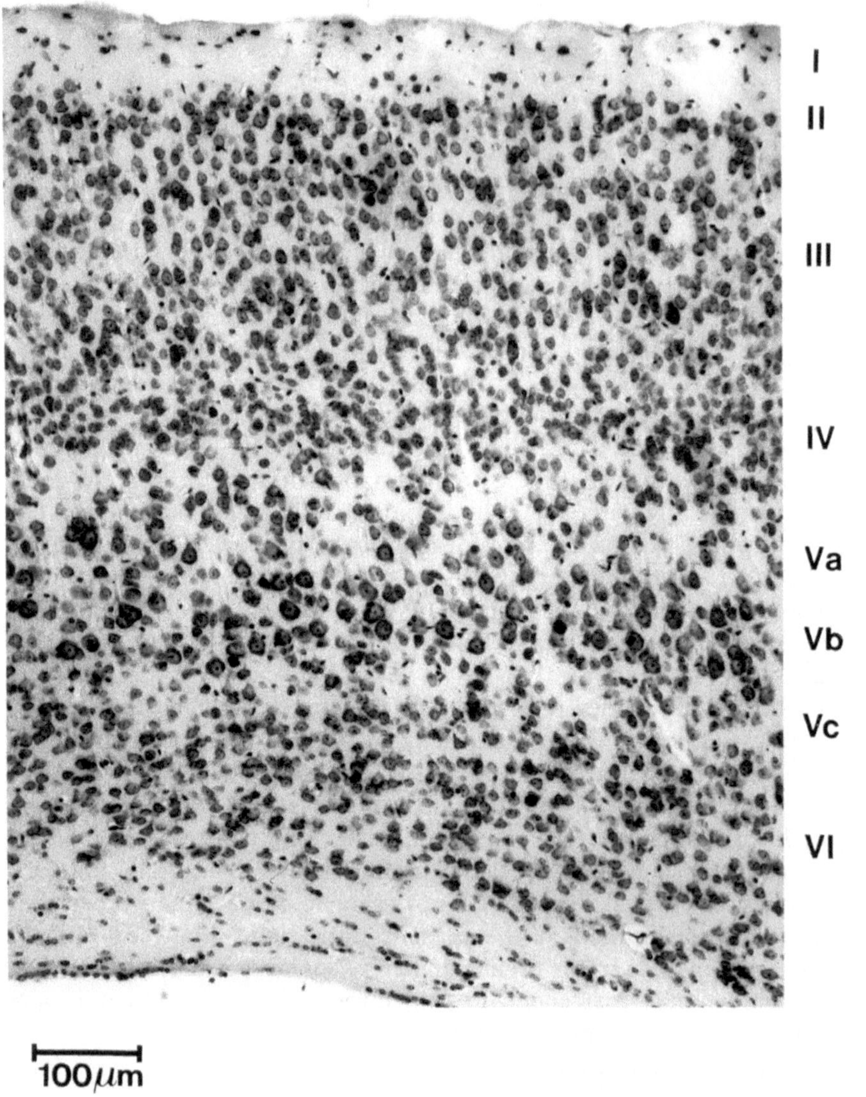

Fig. 63. Layers of the visual cortex (mouse area 17). Nissl stain. *Layers II* and *III* can hardly be distinguished. *Layer IV* contains smaller cell bodies. *Layer V* is subdivided in three sublayers, the middle one conspicuous with its large cell bodies. *Layer VI* is somewhat denser than *layer Vc*

inward from the surface of the cortex, from next to nothing to over 100 000 perikarya/mm^3. For most cortical fields this transition is about 100 μm below the surface. The transition between the first and the second cortical layer which is marked by this density change is equally prominent in marginal regions of the cortex such as the entorhinal area and

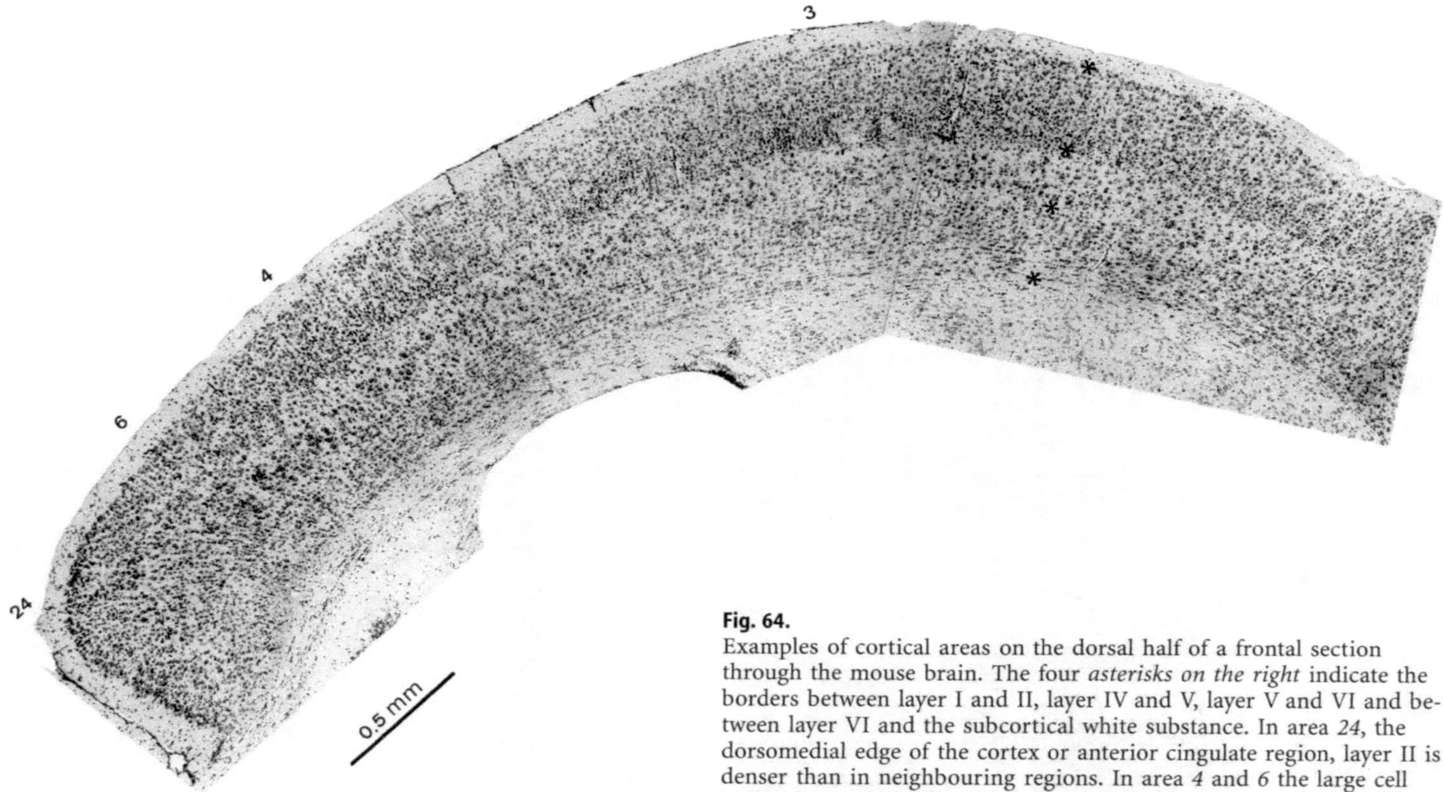

Fig. 64.
Examples of cortical areas on the dorsal half of a frontal section through the mouse brain. The four *asterisks on the right* indicate the borders between layer I and II, layer IV and V, layer V and VI and between layer VI and the subcortical white substance. In area *24*, the dorsomedial edge of the cortex or anterior cingulate region, layer II is denser than in neighbouring regions. In area *4* and *6* the large cell bodies in the fifth layer can be seen even at this low magnification. Area *3*, the barrel field can be easily recognized because of the periodic variations of density in layer IV

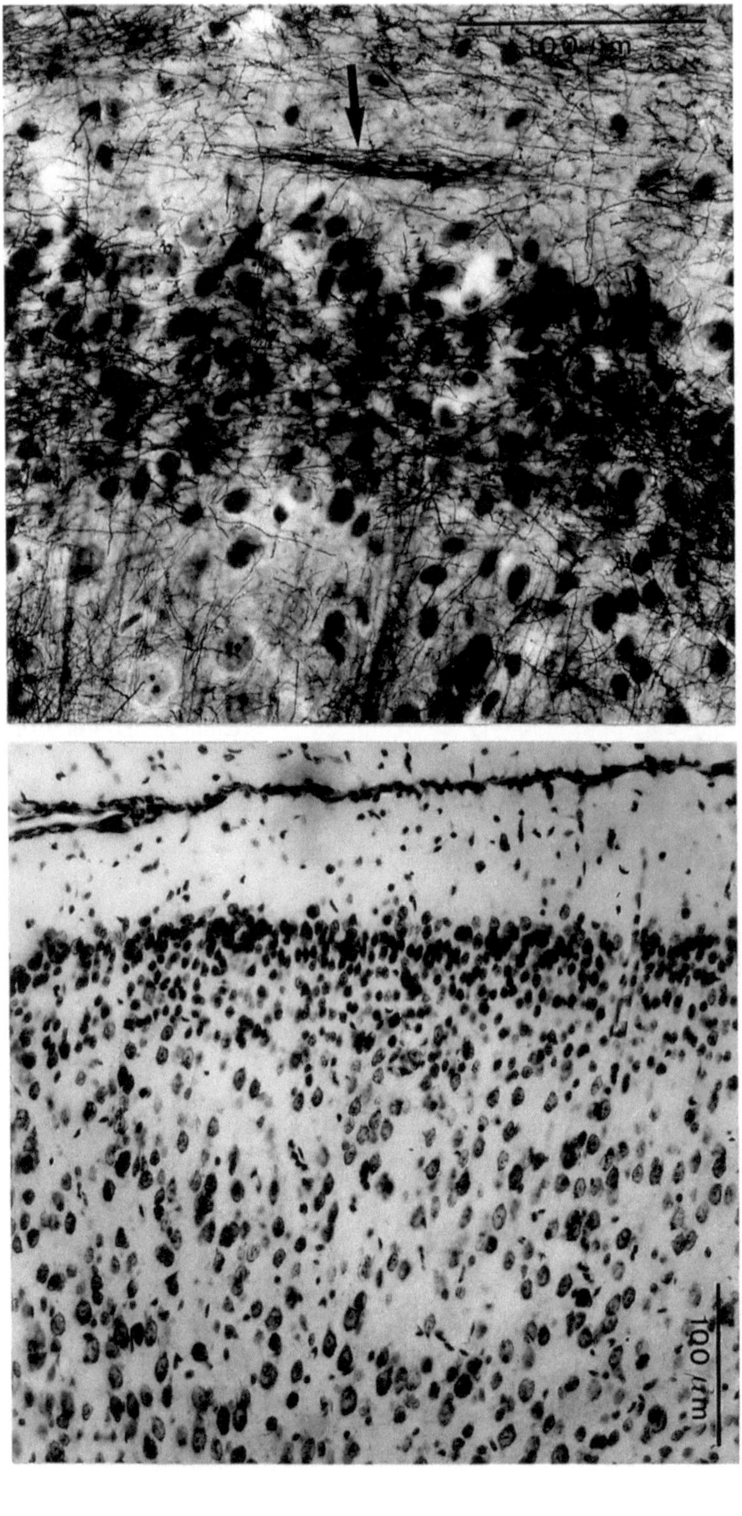

Fig. 65

the subiculum. Following it into the hippocampus (Fig. 75), we find it at a much lower level, corresponding to the upper margin of the layer of hippocampal pyramidal cells. If we take the definition of the first layer literally as the layer above the sudden decrease of the density of cell bodies, the upper two thirds of the hippocampal cortex, comprising the stratum moleculare, lacunare and radiatum all belong to the first layer or rather, are in some way homologous to the first layer of isocortical areas. It remains to be seen whether this homology makes sense also in terms of the specific connectivity of the layers.

The Border Between Layers II and III. In many parts of the isocortex in the mouse this looks like an unnecessary distinction, II and III forming a continuum with the difference in size and shape between superficial cells (layer II) and deeper cells (layer III) easily explainable as a consequence of the shorter apical dendrite of the former. However:

1. In parts of the marginal cortex, II is much denser than III. Most of the cortex ventral to the rhinal fissure, which is to say roughly below the equator of the hemisphere, is characterized by a dense cell layer immediately adjoining layer I (Figs. 3, 4). This is true for the pyriform cortex in the forward part of the brain as well as for the entorhinal cortex in the back (Figs. 2, 75). It is very difficult in the mouse, and we may suppose in most rodents, to homologize this layer either with layer II, or with layer II and III or perhaps even II, III and IV of adjoining regions, because of the discontinuity introduced into the architectonic pattern by the rhinal fissure. Only near the front end of the rhinal fissure does one gain the impression that the superficial dense cell layer of the pyriform cortex goes over into a superficial, also somewhat dense layer above the fissure (Fig. 3), which here can be quite easily identified by horizontal homology with the superficial part of the layer II and III complex, in other words, if one wishes, with layer II.

2. In the posterior cingulate cortex, a strikingly extravagant area (Caviness' area 29b), within the band II-III an upper layer with larger cell bodies may be called II, while III is populated by a multitude of particularly small neurons (Figs. 4, 65). In the Golgi picture both look like pyramidal cells, but with differences in the spininess of their dendrites.

We may conclude tentatively that layer II is not just a hangover from early days of cortical architectonics, when there were two "granular" layers in the

Fig. 65. The characteristic upper layers II and III which give area 29b of the Caviness map (see Fig. 74), the posterior cingulate cortex, its unique appearance. The cells in layers *II/III*, very small in III, slightly larger in II, form a dense band in the Nissl picture *(below)* which coincides almost exactly with a dense layer of fibres in the silver preparation above. (The enlargement is greater in the fibre picture above than in the cell picture below). The fibre bundles in the first layer *(arrow, above)* are another peculiar feature of this area

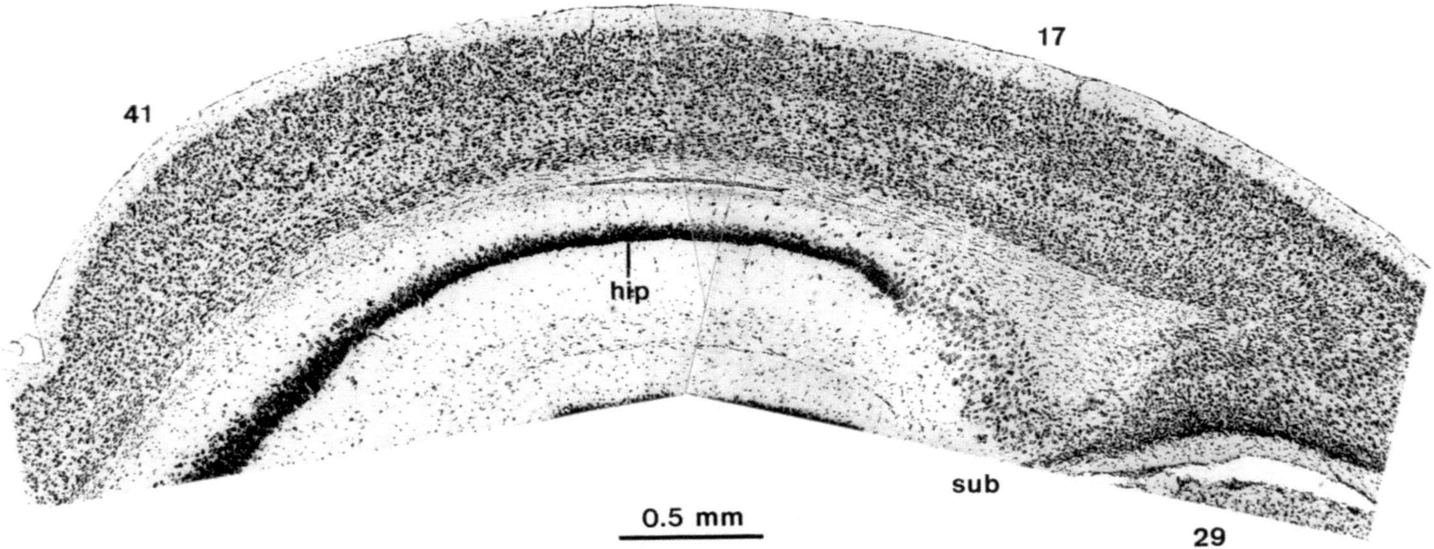

Fig. 66. Architectonic areas on a frontal section posterior to that of Fig. 64 (Nissl preparation). In area *17*, the primary visual area, layers II and III have a columnar appearance. Layer IV is neatly set off from Layer V, which is tripartite: a darker intermediate layer flanked by two lighter layers above and below. The acoustic area *41* is less spectacular in its layering. *hip* the pyramidal layer of the hippocampus; *sub* subiculum; *29* retrosplenial cortex

Nissl picture, the outer one corresponding to the small cells in the superficial tiers of the layer II-layer III complex. Rather, the occasional distinct individuality of layer II leads one to suspect a different connectivity for the two layers, as has indeed been suggested (Jones 1984a).

The Border Between Layer III and IV. This is not, generally, easily visible at low magnification in the mouse cortex. The situation is very different in the human cortex where the many small cells populating layer IV are neatly set off, in many areas, from the generally large and widely spaced pyramidal cells of the lower tiers of layer III. However, by careful inspection at higher magnification one realizes that the same characteristics that mark the transition from layer III to layer IV in the human are valid also in some parts of the mouse cortex, only less strikingly so (Fig. 63). Below a certain level a large number of densely packed smaller cell bodies make their appearance, making the layer below not necessarily darker in the Nissl picture, but different in texture. We have no difficulty in identifying this level with the transition between layer III and IV.

In some places, layer IV is set off by other, more conspicuous characteristics. The periodic variations of the density of perikarya in part of the somatosensory area (Fig. 64), which had been known for some time (Lorente de Nó, 1922; Woolsey, 1967) and were recently identified as the locus of whisker representation and publicized as the "barrel field" (Woolsey and van der Loos, 1970) are limited to layer IV, while layer III seems largely unaffected (at least at the low magnification of Nissl architectonics).

In the largest part of the cortex, on the other hand, we were unable to identify any special layer between the "outer pyramidal cells" of layer III and the "inner" ones of layer V either in the Nissl or in the myelin preparations, at high or at low magnification. It may be that here the lack of a technique which would combine the virtues of the Nissl and the Golgi stain makes itself felt especially.

The Border Between Layer IV and V. Everywhere within a large central region of the hemisphere, at about half-height in the cortex there is an abrupt transition between a layer densely populated with neurons above, and one with fewer neurons below (Fig. 64). This we take by general agreement as indicating the transition from layer IV to layer V. The size of the cell bodies is another indicator, those of layer V being the larger ones (Fig. 63).

When we follow the transformations of the cortical layering that occur in the transition from the central region of the cortex to the hippocampal formation, we notice that this border between IV and V plays a special role by marking the distinction of an upper and a lower main layer of the cortex, the hippocampal formation being derived from the latter (Fig. 75). We shall return to this point in more detail.

Sublayers Within Layer V: the Border Between the Upper and Middle Sublayer. The layer of the cortex which we might call layer V simply because it is situated between the rather well-defined layers IV and VI of the conventional terminology, is in itself not homogeneous.

Again most evidently in the central part of the hemisphere, an upper light sublayer is neatly set off from a darker sublayer below (Figs. 63, 66). The different darkness as it appears in Nissl pictures at low magnification is due to two factors, the higher density of the large cell bodies in the middle sublayer of layer V being one, the presence of cell-free islands in the upper sublayer the other (Fig. 63). These holes in the Nissl picture have a diameter of about 20 to 30 μm and are missing in the middle sublayer.

Distinction of a Middle and Lower Sublayer of Layer V. This distinction can be made in many regions, although less clearly than that between the upper and middle sublayer. Here again, the lower sublayer is the lighter one, contains smaller perikarya and some cell-free islands (Figs. 63, 66). These are less prominent than in the upper sublayer.

The Border Between Layer V and VI. This is again quite well defined in many regions where the lowermost layer of the cortex contains a much denser population of small neurons than layer V does. Moreover, in the Nissl picture, over wide regions of the cortex layer VI is characterized by a horizontal striation (Fig. 64) which is produced by alternating layers of fibre bundles and of cell bodies. There is no such striation in layer V.

28.1 How Real Are the Layers?

Having observed how for each pair of adjacent layers of the classical six (or seven)-layered scheme there is some area of the cortex where the two layers are neatly set off from each other, we have to discuss two questions. First, is it legitimate to follow a layer from one region where it is well defined, into a neighbouring region, where one would not have been able to isolate that layer from its neighbours? Second, how independent are these layers physiologically from each other? The latter question is related to the concept of columns or modules, so dear to some physiologists, and will not be discussed here. The question of the continuity of the layers across areal boundaries, however, is one that touches the core of the architectonic matter and should not be brushed aside. If the cortex were of the same thickness everywhere, and if a boundary between two layers, say IV and V, wherever visible were always at the same height, then there would be no problem. One would rightly suppose that the boundary is at the same level of the cortex even in places where it is not clearly demarcated histologically. But this is not so. The thickness of the cortex varies considerably, as we have already noted,

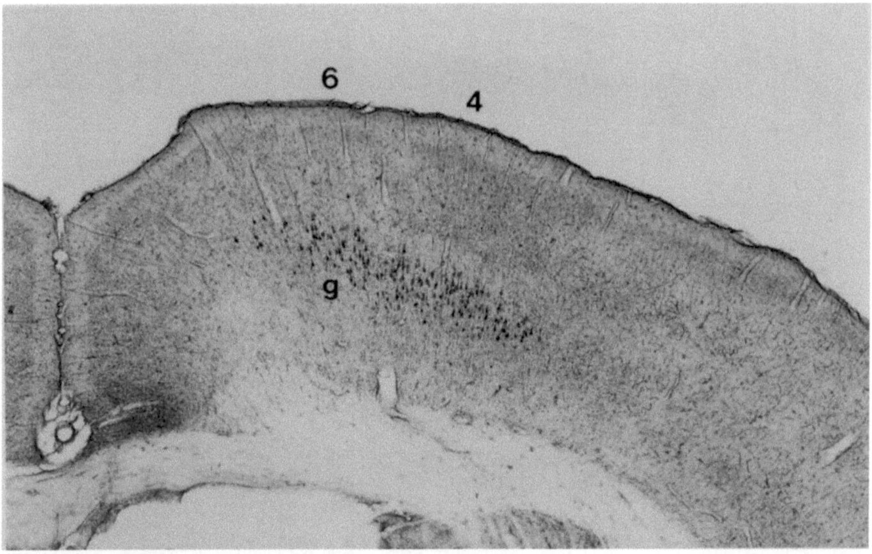

Fig. 67. Correspondence, or lack of correspondence, of the areas defined by their projection and the areas of cortical architectonics. Frontal section through the dorsal part of the right cerebral hemisphere of the rat, at the level of areas *4* and *6*. Injection of horseradish peroxidase into the spinal cord, which marks the cell bodies of the corticospinal neurons, the "giant pyramids" *(g)*. The transition between areas 4 and 6 can be recognized because of the different level of the big cells in layer V (cf. Fig. 64), but the corticospinal neurons are localized in both fields

and the individual layers also occupy various proportions of the cortical thickness. Thus necessarily in some places the borders between adjacent layers run obliquely with respect to the coordinates of the cortex, and where the areal boundary is sharp, (as in the famous example of the 17/18 boundary in primates, but in some places of the mouse cortex as well) we find sudden jumps in the level of one and the same layer as one follows it across the areal boundary. Or is it the same layer? We probably have to wait until we know more about cortical physiology before we can answer this question.

Obliquely running layers or boundaries between layers are not uncommon. For instance, when we follow the band of large cells in layer V in a medial direction starting from an area which we shall later identify as area 4, we observe that the large cells move to a higher and higher position, from well below the middle height of the cortex to clearly above (Figs. 64, 67). This change in position is accompanied by a reduction of the thickness of layers I to IV and by a corresponding increase of layer VI.

Similarly, at the level of the rhinal fissure, in a large part of the hemisphere, the layers II to IV may be considered to merge into the dense

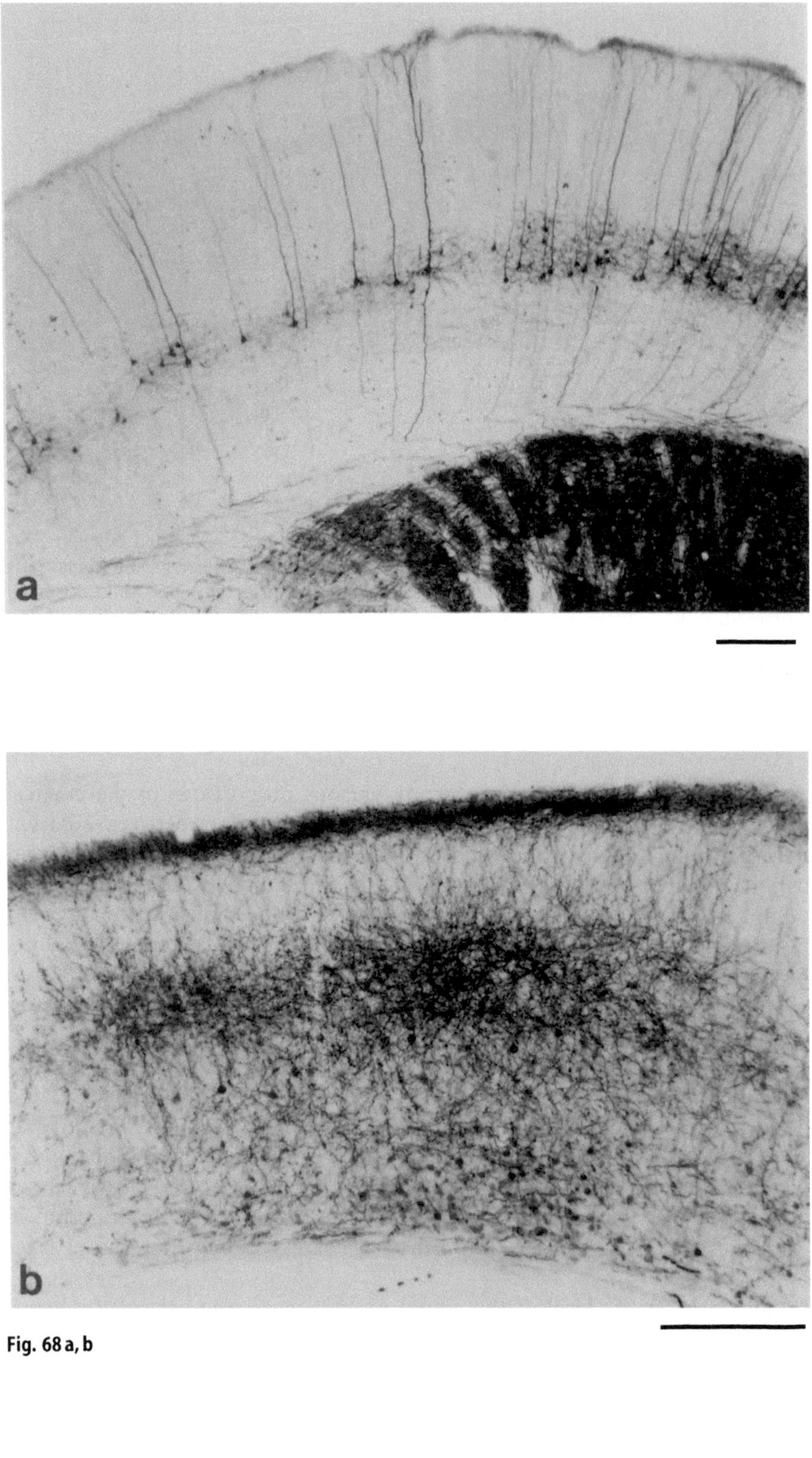

Fig. 68 a, b

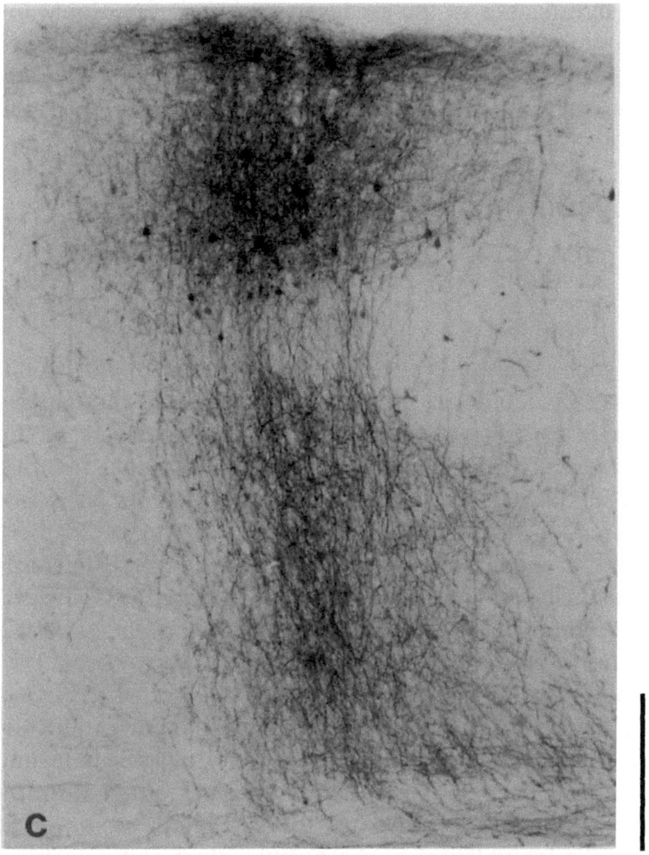

Fig. 68 a–c. **a** Retrograde staining of pyramidal cells in layer V after injection of biotinylated dextran amine (BDA) into the descending pathway (Crus cerebri). **b** Afferent thalamic fibers (primary acoustic cortex, area 41 on Fig. 74) revealed by anterograde transport after injection of BDA into the medial geniculate body. **c** Termination of callosal fibers stained after injection of BDA into the opposite hemisphere. Note the separate localization of these elements in different cortical layers. Bars: 200 μm

superficial band, which characterizes most of the cortex of the ventral hemisphere (Figs. 3, 4). However, as we have already mentioned, at least in the rostral part of the pyriform cortex, an alternative interpretation of this cell layer as being just layer II, with a separate layer III underneath, is also possible.

We do have better definitions of the layers now in terms of external and internal connectivity (Jones, 1984a; White 1989). The thalamic input, for example, is mainly relayed to layer IV (Fig. 68b), and the cortical output to distant parts of the brain comes mainly from neurons in layer V (Figs. 67 and 68a). The terminations of callosal fibers are distributed in upper and lower layers of the cortex and appear to avoid layer IV where thalamic afferents mostly terminate (Fig. 68c).

On the whole, the significance of the six layers and their functional interplay remains a mystery awaiting theoretical treatment. A recent hypothesis by Miller (1996), based on differences in both connectivity patterns and spontaneous activity between upper and lower layers, assigns to the upper layers the role of a neuronal "library", storing most of the information encoded by assemblies (see chapter 35), while the lower layers are assumed to *catalyze* the process of assembly formation.

28.2 Aberrant Layering

(If it may be called such). We have stretched our method of horizontal continuation across areal boundaries to the point where we thought we could identify one layer as the result of two or even three layers of the six-layered scheme merging with complete loss of their original identity. The dense cell band of the pyriform cortex was mentioned as an example. There are other cases in the marginal regions of the hemisphere where one layer seems to stop abruptly, while for other layers the continuity can be clearly established. The sixth layer, which is rather conspicuous everywhere in the central part of the hemisphere, disappears within a small transitional zone when one tries to follow it into the pyriform cortex (Figs. 3, 4). Another striking case of the disappearance of a cortical layer can be seen in the transition between the entorhinal cortex and the subiculum (Fig. 75). There, at the border between the presubiculum and the subiculum the entire upper dense layer of the cortex stops abruptly, while the simple, homogeneous cell layer of the subiculum seems to be derived entirely from the lower layers of the presubiculum and entorhinal cortex.

But there are also places where additional layers make their appearance. Where the entorhinal cortex borders the more rostral regions of the surrounding cortex, within the entorhinal, an additional superficial cell layer is seen above layer II. This splitting of layer II stops abruptly especially in a rostral direction.

Chapter 29

Cyto- and Myeloarchitectonics: Two Aspects of the same Reality

What leaps to the eye in the Nissl picture viewed at low magnification is the alternation of "darker" and "lighter" layers. At a higher magnification it becomes obvious that the darkness is composed out of two parameters, number and size of neural cell bodies. (We neglect another possible source of variation, the density of Nissl bodies within the perikaryon.) How can we interpret these two variables in terms of connectivity?

The size of the perikaryon is correlated with the volume of the whole neuron, i.e. with the size of its dendritic plus axonal expansions. Thus we expect *large cell bodies* when the axons leaving the cortex are very large, such as is the case of the giant pyramidal cells of the motor area. But apart from these, the size of the cell body reflects mainly the size of the dendritic tree (Bok, 1959). We expect and in fact often find lighter zones flanking a layer which contains large perikarya. The lightness is due to the large dendrites of these neurons which occupy much space but being unstained do not contribute to the darkness of the Nissl picture. We would be very surprised to find a layer or region where there is a low density of *small perikarya*: it would not be clear where the cell processes should come from which fill the space between them. A dense layer of small perikarya may, however, be adjoining a very light layer, if all the dendrites of these small cells collect in a neighbouring layer.

Summarizing, we expect the following situations:

- Large neurons, relatively high density: this indicates long efferent axons. The most striking example is the motor area of larger animals, but a similar situation can be seen in Area 6 of the mouse (Figs. 64 and 67).
- Large neurons in a low density region: this is an indication of large or very densely ramified dendritic trees. In the mouse, layer V in many areas is an example for this.
- Small neurons, high density throughout: these must all have small, well interwoven dendritic trees. We have this situation in layer IV of sensory areas, e.g. in Area 17 (Figs. 63 and 66).
- High density of neurons but flanked by lighter regions (layers): this is an indication of an oriented, parallel arrangement of dendrites. The most striking example is the hippocampus. Also layer II/III of the posterior cingulate cortex (Fig. 65) and the barrels of Area 3 (Fig. 64) can be explained in this way.

In addition to the distribution of dendrites, also intracortical axon bundles may influence the Nissl picture. In many areas neural cell bodies are arranged in vertical chains, apparently constrained by bundles of afferent and efferent fibres.

On the whole, however, the density, size and distribution of nerve cell bodies in the layers and areas of the cortex offers only tenuous clues to a functional understanding. A complete repertory of neuronal shapes, as revealed by Golgi and similar methods could provide much more significant information on the local variations of cortical "wiring" (statistical wiring, that is, in our philosophy). However, the available surveys are lacunary (Cajal, 1911, from which our Figs. 84 and 85 were taken, is still the best) and sorely lacking in quantitative assessments. Quite apart from the unreliable nature of the Golgi staining procedure, the measurements of axonal and dendritic trees of individual neurons, as exemplified in this book, are too laborious to be applied to a large population of neurons.

One would like to use a method which directly provides a statistical picture of connectivity, such as the silver stain of our Fig. 20 in which the entire axonal felt is stained. However, in the thin sections required for this technique it is hardly possible to recover the relevant information referring to the 3-D network, such as the length and direction of all axonal branches.

Myelin preparations are an alternative. They show only a fraction of all fibers and therefore may be cut into much thicker slices without losing their transparency. Moreover, the fibers stained being the myelinated ones, and for the most part of greater than average caliber and length, one may expect to see in the myelin picture something like a more macroscopic, simplified version of cortical connectivity as it varies in different areas. The myelinated fibers very probably do not contribute to the "plasticity" of the cortical network for which we hold changes in synaptic strength responsible, since they carry no synapses. They may, rather, be taken as embodying the constraints which antedate the learning mechanism, as the cables that bring information together in the right places where it may be incorporated in the synaptic memory trace. Of course, the preexisting structure of the "learning matrix" embodied in the myelinated axons may in itself be the outcome of early experience, if the formation of the myelin sheath, as has sometimes been suggested, depends in some way on activity. The function of the myelin may be to insulate those axonal segments once and for all, which in an early phase had failed to establish functional relations with neighbouring dendrites, perhaps because nothing worth remembering was discovered there.

Unfortunately, in the mouse cortex myelinated fibers are scarce, not enough to describe variations of cortical connectivity in terms of their distribution. They are very conspicuous in larger animals, notably in the cat and in primates. For this reason we shall temporarily abandon the mouse in this chapter and base our comparative analysis of cyto- and myeloarchitectonics on the human brain, where the areal variation of cortical structure is known in great detail.

Special patterns of myelination were discovered early in the human brain (the "striate area", Gennari, 1782, the "stripes of Baillarger" of some areas, Baillarger, 1840). They became the subject of systematic investigations in the hands of Vogt (1910, 1911) and of some of his followers (Hopf, 1954, 1955; Sanides, 1962). However, the maps and the numbering of human cortical areas which found the favour of practising neurophysiologists and clinicians were not those obtained from myelin preparations, but from sections stained for cell bodies with the Nissl technique (Brodmann, 1909 whose numbering of the areas, even if curiously illogical, was universally adopted; v. Economo and Koskinas, 1925, who provided an impressive pictorial documentation). The myelin picture was occasionally disinterred, side by side with the Nissl picture (Fig. 69), but only to illustrate the apparently unsystematic discrepancy between the cortical layers in one and the other kind of preparation. To mention just one example, the two maxima of myelinated fiber density that go by the name of "strias of Baillarger" in the myelin picture, are situated at levels in the cortical thickness corresponding, in the Nissl picture, to a dense layer of small ("granular") cell bodies in one case, to a loose layer of large cells in the other.

The two kinds of preparation certainly reflect two sides of the same reality. On several occasions (Braitenberg, 1962, 1974a, 1978b) we turned to the problem of the rules relating the distribution of cell bodies and fibers in the cortex, until finally Hellwig (1993) provided a very convincing explanation based on a few simple assumptions and corroborated by quantitative analysis. A summary of this particularly impressive piece of theoretical neuroanatomy follows.

Since the distribution of cell bodies and the distribution of myelinated fibers are subject to variations over the surface of the cortex, it is natural to assume that both reflect the same local adaptations of the basic cortical machinery required by different functional specializations. If this is so, we expect the areal boundaries on the cytoarchitectonic map to coincide with the boundaries drawn on the basis of myeloarchitectonics, which is true in most cases, as Brodmann (1909) had already observed. Of course, it would be desirable to know *in what way* the basic cortical network is adapted locally in order to comply to the various functional needs. If we knew the essential features of the local variations in terms of differences in the convergence/divergence factor between neurons, of the relative importance of their long and short range connections, of the strength and range of inhibitory interneurons, it would probably be possible to predict the areal variations as they appear in the Nissl and in the myelin preparations. As it is, we have no theory by means of which we could derive the optimal statistics of the network in the context of visual or acoustic perception, in the motor or language areas of the cortex. Pro-

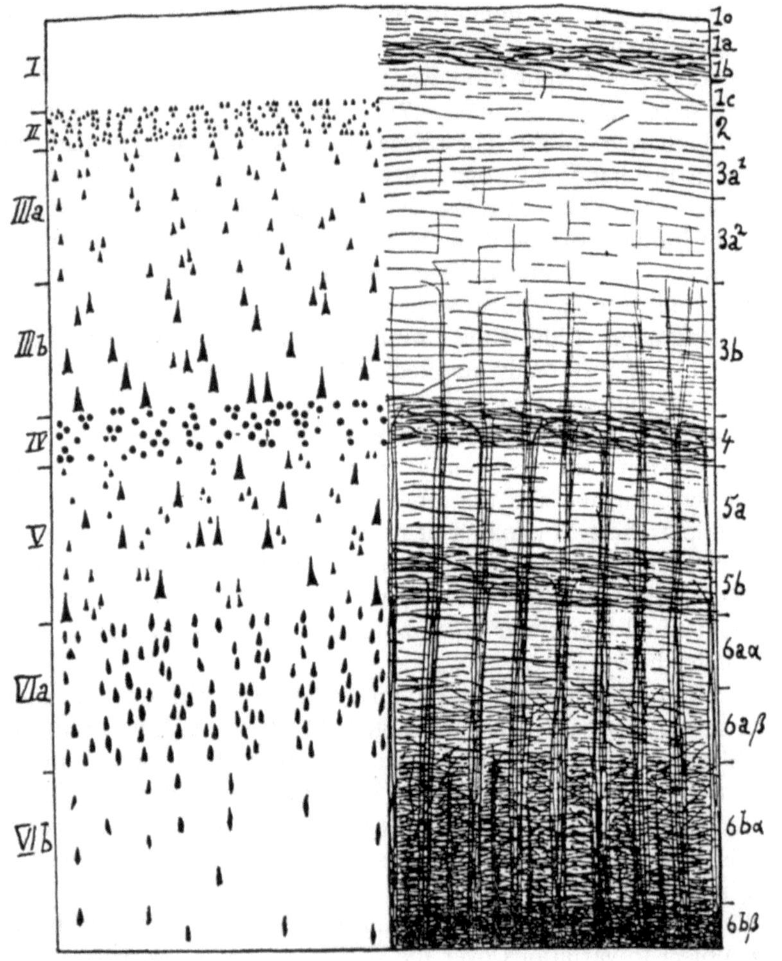

Fig. 69. Correspondance of the layers as seen in Nissl preparations with those in the myelin preparation, according to Oscar Vogt. The illustration is taken from Bailey and von Bonin (1951)

ceeding in the opposite direction, however, we can hope to come close to an understanding of these local functional adaptations by inferring their essence from architectonic areal variations. But first we must determine the relative merits of the cyto- and myelin-picture as clues to the cortical statistics, and how they are related.

The main assumption made by Hellwig (1993) following an earlier suggestion (Braitenberg, 1962) was that the population of intracortical myelinated fibers consists mainly of the myelinated segments of axons and axon collaterals of cortical pyramidal cells, with only a minor contribution from ascending extracortical afferents (mostly of thalamic origin) and even less from inhibitory interneurons. This led to the assumption that the amount of myelin in different layers depends on the shape, size and distribution of pyramidal cells in the cortex. The next step was to find empirically a rule relating the amount of myelin contributed by a single pyramidal cell to the size of its cell body, and the distribution of the myelin to the localization of the cell in the cortex. It turned out that two simple rules were adequate: (a) the amount of intracortical myelin contributed by each pyramidal cell is a monotonically increasing function of the size of its cell body, described by a sigmoid curve; (b) the myelinated fibers belonging to a single pyramidal cell are distributed according to a density function which has a marked maximum about 0.2 mm below the level of the cell body and tapers off gradually toward the bottom and more steeply toward the surface of the cortex. Both these assumptions were derived from histological observations on Golgi preparations. While no actual measurements were available for the first, data for the second, referring to the location of axon collaterals relative to the cell body, were mostly derived from rodents and had to be scaled to the different thickness of the human cortex.

Ultimately, the success of Hellwig's method justified the correctness of his assumptions. He showed that the sum of the individual contributions to the network of myelinated fibers made by all the pyramidal cells of a piece of cortex, when calculated according to the assumptions (a) and (b), produced a density profile of myelin remarkably similar to the one described in myeloarchitectonics. In other words, he could predict for each area the pattern of layering in the myelin preparation, given the density of large and small (pyramidal) cell bodies in the different layers as seen in the Nissl picture.

The most exhaustive survey of human cortical cytoarchitectonics was provided by v. Economo and Koskinas in 1925. This monumental work includes an atlas of large microphotographs of all cortical areas as they appear on Nissl-preparations cut at right angles to the cortical surface. Most importantly, it also specifies for each area precisely those quantitative assessments which Hellwig needed, namely thickness of the cortex, thickness of the layers, number of cell bodies and their average size for each sublayer.

Hellwig calculated the myelin density profiles from the v. Economo and Koskinas data for 14 cortical areas, including the primary sensory areas, the motor cortex, some frontal, temporal and parietal so-called associational areas and the frontal and temporal language centers. In each case the theoretical density profile of the myelin calculated fom the cytoarchitectonic picture matched the actual myelin picture of the same area. In areas where the outer stria of Baillarger is darker than the inner one (e.g. the postcentral Area 2, temporal Area 22), the simulated myelin picture showed the same characteristic. In other areas (e.g. postcentral Area 3, superior temporal Area 41) the inner stria is more prominent, and again the same is true in the simulation. Even in areas which appear very different from all the others in the myelin preparation, such as the temporal pole (Area 38), which is very poorly myelinated, or the primary visual cortex (Area 17) with its single prominent maximum of myelin in the "stria of Gennari", the prediction of the myelin picture based on cytoarchitectonic data was astonishingly accurate.

In Fig. 70, from Hellwig (1993) the myelin pictures (d) of three cortical areas (the parietal Area 7, the occipital Area 17 and the precentral Area 4) are shown together with the corresponding Nissl picture (a), with the computed myelin density as a function of cortical depth (b) and with the simulated myelin picture (c, which is the transformation of the graph shown in b into shades of grey).

The success of Hellwig's Nissl-to-myelin transformation warrants some comments related to the general theme of our book.

1) Since Hellwig considered only pyramidal cells, the impression one gains is that local areal specializations principally involve the structure of the *skeleton cortex of pyramidal cells.* It may, of course, be that the number, size, form and distribution in different layers of non pyramidal cells also vary in different areas, and sporadic observations to that effect (e.g. Lund et al., 1981, referring to the distribution of *chandelier cells* in the visual cortex) have been reported. To us, however, the dependence of the size and number of pyramidal cells on the functional specialization, and presumably on the kind of input reaching the cortex in different areas strongly suggests that the specific adaptations of the cortex essentially apply to the pattern of connections between pyramidal cells. In other words, the statistics and the geometry of the pyramidal-to-pyramidal cell connections which antedate learning, are optimally adapted to the statistics and the geometry of the activity patterns with which they will be confronted. Assuming that in one sensory field most of the correlations that will be discovered and learned refer to short range interactions, while in another sensory field correlations between elements far apart are to be expected, it would be wise to predispose the network in different ways in the two areas, with many short horizontal fibers in one case, and as many as possible long fibers in the other. If this is the principle, inborn knowledge is essentially statistical in nature, or, to put it another way, refers to the redundancy of the expected signal flow in various sensory contexts.

2) The fact that Hellwig's study considers only myelinated fibers may not be a handicap. A critic could easily maintain that the myelin picture, far from showing the pattern of connections between neurons, actually shows where connections cannot be found, since the fibers isolated by myelin sheaths can hardly make any synapses. However, this is in itself relevant information, showing, as it does, something

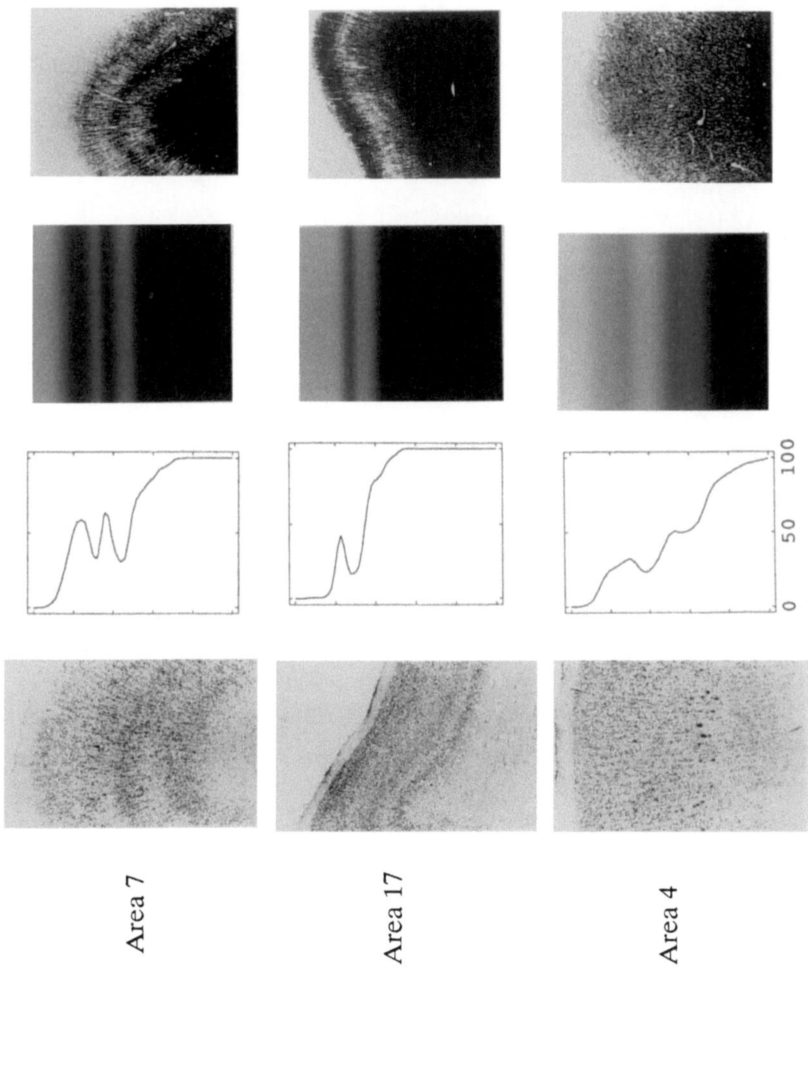

Fig. 70 a–d. From Nissl pictures of the human cortex (**a**) the density of myelin fibres in different layers can be calculated (**b**) and graphically displayed (**c**). The resulting distribution is strikingly similar to real myelin preparations of the same areas (**d**): parietal area 7, occipital area 17 and precentral area 4. The graphs in b display the density of myelin fibres in percent of maximum on the horizontal axis, the depth in the cortex on the vertical axis. Modified from Hellwig (1993)

like the negative of cortical connectivity. Besides, if all cortical fibers are stained by a good reduced silver method, such as the one used for our Figs. 1 and 20, the overall distribution of fibers in the cortical layers is very similar to that in the myelin preparation, only much denser. In all animals, the number of myelinated fibers is only a small fraction of all axonal fibers, a fact for which we have no explanation at present.

Chapter 30

Special Areas

We shall now describe such cortical fields as easily leap to the eye, even to a superficial observer. We notice, however, that there are two different situations that may make a delimited region of cortex clearly stand out between its neighbours. An area may be defined by a boundary where the appearance of the layering abruptly changes. The classical example in human anatomy is the striate area (area 17, Brodmann). There are other areas, however, which equally leap to the eye, because of prominent characteristics, even if they may blend gradually into the appearance of surrounding areas. A well-known example for this is the giganto-pyramidal area, area 4, of the human cortex, which is well characterized by the enormous cell bodies of the Betz cells, but much less by its boundaries. We find examples for both in the mouse cortex. In the numbering of areas we shall follow Caviness (1975).

30.1 Areas with Sharp Boundaries, Almost Everywhere

Posterior Cingulate Cortex (Retrosplenial Cortex, Area 29b). This is a remarkable field, easily recognized, both in the myelin and Nissl preparation (Fig. 65). In the cell picture the prevailing characteristic is a band of very small cell bodies occupying a level between 70 and 160 μm from the surface. This corresponds to the levels of the second and third layer. What is remarkable is the dense packing, especially in layer II. These small cell bodies belong to tiny neurons which have all the characteristics of pyramidal cells in the Golgi picture. The distinctive feature of this field in the myelin and reduced-silver preparation is a dense layer of fibres almost exactly coincident with the band of small cell bodies characteristic of the Nissl preparation (Fig. 65). Only the uppermost level of the dense layer of small cells, which we have identified in a previous chapter as layer II, appears to lie above the dense fibre layer. Like the cell band, the fibre band also comes to an abrupt end at the margin of the field.

Another characteristic of this region is a rather conspicuous population of fibres, especially in the upper part of layer I (Fig. 65 above) all running in roughly the same (dorso-ventral) direction and forming distinct bundles in some places.

We found it impossible to identify something like layer IV in this area of the cortex. Immediately below the layer of the cell bodies of the dwarf pyramids there is a transition to a layer of widely spaced, fairly large cells, most likely of the pyramidal kind (Fig. 65 below). This layer has all those characteristics which we usually associate with layer V.

For an extensive account of the retrosplenial cortex see Vogt (1985), Michalski et al. (1979), Herrmann and Schulz (1978), van Groen et al. (1993).

The Barrel Field. This field appears, at first sight, to have every characteristic of a well-defined cortical area: a very special appearance at low magnification (Fig. 64) and sharp boundaries. The defining trait is a periodicity of the picture of the fourth layer obtained in tangential Nissl-stained sections. The honeycomb-like appearance is due to stretches of high cell density surrounding islands of lower density. The islands are clearly arranged in rows and columns, although the rows are bent and the size of the periodic subunits varies over the entire region. The cytoarchitectural peculiarities of this area are well described in Woolsey (1967) and Woolsey and van der Loos (1970), where we learn that the "barrels" have a diameter varying between 100 and 400 μm (we would have set the upper limit a little lower) and that there are about 200 of them in the whole field.

We also learn that the posterior part of the field containing the larger barrels is neatly set off from the surrounding regions, where the fourth layer suddenly becomes homogeneous again. But the anterior half of the field is not, the periodicity of the barrels being gradually submerged in the continuous structure of the fourth layer elsewhere. One wonders if careful analysis would not reveal a sharp boundary around the frontal region of the barrel field as well. This is a point of general importance, since the gradual disappearance of a very obvious "columnar" arrangement would lead one to suspect that the rest of the cortex is columnar too, only less markedly so. On the other hand, if the periodicity stops exactly at the margin of the cortical field, we would rather put the pattern in relation with the periodic structure of the thing represented, quite independently of any general principle of cortical organization. And here it is indeed obvious that the discreteness of cortical representation repeats the discreteness of the vibrissae on the muzzle of the mouse (Woolsey and van der Loos, 1970) and therefore does not necessarily reflect an a priori tendency of the cortex to segregate its sensory input into discrete, homogeneous units.

This is not to say that there is no such tendency in the rodent cortex. In the rat, it has been shown that barrel-like periodic structures are not restricted to the whisker representation, but seem to be a more general feature of the rodent somatosenory cortex (Welker, 1976; Chapin and Lin, 1990). For example, in the region where the forepaw is represented, barrels (or groups of barrels) may correspond to parts of digits (Waters et al., 1995).

30.2 Areas with Characteristic Layering

The Giganto-Pyramidal Area. In larger mammals, the motor area, or area 4, located on the anterior bank of the central sulcus, is one of the most prominent regional specializations because of the large or "giant" pyramidal cells of the fifth cortical layer. Similarly, in the mouse, a region of the anterior medial part of the convexity is defined by pyramidal cells of larger than usual size in layer V (Fig. 64). Measurements of the average size of cell bodies in that layer generally give slightly larger values than in neighbouring regions. (In Nissl preparations, the perikarya of the cells were drawn and the area of these projections was measured with a digital planimeter). But the largest perikarya, with an area of more than 177 μm^2 (corresponding to circles with a diameter of 15 μm) were found only in a restricted region, which we have chosen to call giganto-pyramidal. This region corresponds roughly to areas 4 and part of 6 of the Caviness map. Clearly, this area is composed of two different regions characterized by different layering and different distributions of cell sizes: in the anterior region (area 6) the large cells of layer V fill the whole layer up to the border of layer IV, which is comparatively loose and thin. In the posterior-lateral region, which we choose to call area 4 (although it turns out to be somewhat larger than indicated on the Caviness map), the large neurons of layer V tend to be clumped and do not reach up to the upper delimitation of that layer. Also, layer IV is denser in area 4. At a lower magnification, area 4 is well characterized by a conspicuous light layer Va. It would be most desirable if the architectonic definition of area 4 were to coincide with that region which sends direct projections to the spinal cord. This can be assessed by injections of HRP into the spinal cord. Figure 67 shows the distribution of cell bodies stained with the HRP-reaction product after injection at the thoracic level of the rat spinal cord. The dense band of cells projecting to that level is wider than area 4 on the Caviness map if we are allowed to compare two rodents of different size. The non-correspondence of the architectonic boundaries with those of the areas defined through their projection can be seen on one and the same section (Fig. 67). The border between the two regions can be seen to cut across the region containing the neurons which project to the spinal cord.

Area 17. One would probably not have singled out that region in the mouse cortex, had physiological evidence not pointed it out as the primary visual field. While they do not hit one in the eye, its morphological features are, nevertheless, evident: there is a dense and wide granular layer IV ending below with an abrupt transition to layer V (Figs. 63, 66). Layer V is clearly subdivided into three sublayers: a middle one containing many large cells flanked by two layers only sparsely populated by cell bodies. The cortex in area 17 is thicker than in neighbouring fields (Fig. 66).

On tangential sections one notices roundish patches devoid of cell bodies in the lower part of layer IV and in layer V.

Primary Acoustic Area. Compared to the somatosensory and visual area, the primary acoustic area (Fig. 66) is not characterized by very marked architectonic traits, except that it is conspicuously thicker than the neighbouring cortex. (This may have something to do with the special role of the acoustic area in the convergence/divergence pattern of cortico-cortical connections, see Chap. 26 and Figs. 60 and 61). The transition between the visual and acoustic areas is, however, quite sharply defined at the border of area 18a and 22 of the Caviness map. Layer IV is much more clearly defined on the visual side of this border and layer V more clearly broken up in sublayers.

30.3 Characteristic Areas in the Lower Part of the Hemisphere

It is convenient to separate the anterior parts of the ventral hemisphere, surrounding the primary olfactory areas, from the posterior parts in the proximity of the hippocampus. We shall start again with some remarkable fields well-defined by sharp boundaries or by unmistakable characteristics. The whole anterior basal cortex is characterized by a high density of cells in the second layer.

Cortex of the Tuberculum Olfactorium (Fig. 3). This, the most anterior part of the basal cortex, is defined by a very striking tendency of groups of small cells to separate from the second layer, which in these places appears to almost reach the surface of the cortex. Groups of similar cells are seen, isolated from the cortex in the very depth of the septal tissue. These strange formations were described by Cajal and termed Insulae Callejae. In Nissl pictures these islands are striking because of the high density of small, darkly stained nuclei. Microscopic observation shows quite clearly that these do not represent a separate cell type, since the very small dark nuclei are seen to change gradually into the ordinary nuclei of the 2nd cortical layer. It may of course be that the Nissl picture hides some fundamental distinctive characteristic, but we have no evidence from our own Golgi material.

The border between the pyriform area and the tuberculum olfactorium (Fig. 3) is also well defined, especially in a more anterior region, by a sudden bend of the plane marking the transition from the first to the second layer. The first layer is much thinner in the Tuberculum olfactorium.

Pyriform Cortex with Claustrum. The SAP atlas marks as claustrum the lower layers of the part of the pyriform cortex which adjoins the lateral isocortex (Figs. 3, 4). On superficial observation this appears to be nothing but the continuation of the isocortical layer VI and perhaps lower layer V. In any case there is no "capsula extrema" of white substance separating the claustrum from an overlying insular cortex, as we see it in larger mammals.

Pyriform Cortex Without Claustrum. In the basal anterior cortex the very thick first layer, very heavy second layer (perhaps II plus III?) and the very loose cell layers below are sufficient definition of the main part of the pyriform cortex (Fig. 3).

In the posterior pyriform cortex the dense cell layer (II or II + III) shows irregular clusters of cell bodies (Fig. 4). This region partly overlaps with what we have termed pyriform with claustrum.

Posterior Basal Cortex. This is practically identical with the entorhinal region and with the cortical amygdaloid nucleus of the SAP atlas. Although it is attractive to consider the entorhinal cortex as the continuation of the pyriform cortex, with the dense entorhinal second layer continuing the characteristic dense layer of the pyriform cortex, we believe that a more organic description can be based on the concept of the hippocampal formation.

The Hippocampal Formation. As is evident on Fig. 2 and Fig. 75, the hippocampus continues the posterior margin of the cortex. If the cortex is seen as triangular or cone-shaped (Fig. 5), the hippocampus is antipodic to the olfactory input. Its position at the margin of the cerebral hemisphere may well be associated with its function. Hence we begin our description with the most marginal strip: the dentate gyrus.

The Fascia Dentata. This piece of cortex occupies a very special position in that the cortical plane of the dentate gyrus (or fascia dentata) does not go over continuously into the plane of the rest of the cortex as we have already mentioned (Figs. 2, 5, 75). The fascia dentata is shaped like a trough containing the margin of the hippocampal cortex in its concavity (Fig. 5). The very simple layering is also remarkable. There is a cell body layer in which small neurons are very densely packed, forming the lower third or fourth of the dentate cortex. The upper "molecular" layer is decidedly lacking in neural cell bodies.

It is debatable whether the somewhat disorderly array of cells in the so-called hilus of the fascia dentata just underlying the dense cell body layer belongs to the dentate fascia as its name tends to suggest. Judging from its connections, it would rather be part of the following region.

The Hippocampus Proper. This is characterized by a remarkable peculiarity in the arrangement of the neurons. The main neural population, the hippocampal pyramidal cells, are very homogeneous in size and reach from the upper to the lower margin of the hippocampal cortex. Their cell bodies are concentrated in a narrow band, in striking contrast with the staggering of neural cell bodies in other parts of the cortex between the 2nd and 6th cortical layer. The Nissl picture would not suggest at first sight the subdivision of the hippocampus proper into two approximately equal halves, which becomes imperative when the fibre connections (in particular the mossy fibres, see Chap. 32) are considered.

Proceeding from the hippocampus toward the centre of the cerebral cortex, we meet a rather abrupt transition into another strip of cortex, well defined by a wider dispersion of cell bodies in the lower cortical layers and otherwise quite homogeneous; the subiculum (Figs. 2, 66, 75). It is quite apparent, especially in the lower posterior part of the hippocampal formation, how the layer of neural cell bodies in the subiculum goes over into the *lower* layers of the adjoining cortex. Hence we may consider the whole subiculum-hippocampus-complex as a protrusion of the lower cortical main layer (Braitenberg and Schüz, 1983). As abrupt as the transition from the hippocampus to the subiculum is that between the subiculum and the adjoining cortex, the so-called entorhinal cortex. As one proceeds from the subiculum, a layer of cell bodies in the upper part of the cortex makes its sudden appearance (Fig. 75).

This upper cell layer bunches up in a peculiar way between the subiculum and the entorhinal cortex giving to some the opportunity to define a transitional area, the presubiculum. The entorhinal area is very easily recognized in the low magnification Nissl picture because of the two addensations of the cell bodies, separated by a lighter intermediate layer, the upper one which starts with the presubiculum and the lower one which continues the subiculum (Figs. 2, 75).

Within the entorhinal area, in its most ventral-posterior part, there is the specialized region defined by the splitting of the upper cell layer into two distinct sublayers already mentioned in Chapter 28.

We shall return to the hippocampus in Chapter 32 from the point of view of its internal "wiring".

Chapter 31

Map of the Cortex

For the purpose of representing areal subdivisions of the cortex in a unitary picture, it seemed desirable to produce a planar map of the surface of the cortex.

It is intuitively obvious that a curved surface can only be represented without distortion on a plane in special cases, namely in the case of a cylindrical or conical surface. In other cases, e.g. in that of a spherical or of a saddle-shaped surface, tears, folds or distortions of the distances represented cannot be avoided, as cartographers well know. (This has something to do with the following observation: on the mantle of a cylinder or of a cone, as on a plane, the locus of all points having a distance r from a given point is a closed line of length $2\pi r$. On the surface of a sphere this length is less, on a saddle more than $2\pi r$.)

Luckily, a large part of the mouse cortex can be approximated by the surface of a cone having its tip near the olfactory bulb and its base opposed to the cerebellum and mesencephalic tectum (Figs. 2, 5). The cone is open ventromedially, and the upper part of the medial wall is flattened into the interhemispheric cortex, thereby constituting no obstacle to ironing out in a plane. Along a cut running in an antero-medio-dorsal to postero-latero-ventral direction, the surface of the cone goes over into the convoluted surface of the hippocampus (Fig. 5). This, however, is to a good approximation a surface of a cylindrical type and can be unrolled and flattened out without major distortion.

The only two regions of the mouse cortex where definite bulges resist flattening into a planar map are the frontal and the occipital pole. It should be noted that these regions appear too small on our map, and the surrounding regions somewhat too large, compared to the rest of the cortex.

Fig. 71 was constructed in the following way. The length of the cortex was measured, between the borders already described in Chapter 2, on standard sections chosen from the sagittal, horizontal and frontal sections of the SAP atlas. The sagittal section 119 was drawn as a straight line and, forming a certain angle with it, the horizontal section 101 also. The angle was determined thus: characteristic cortical places, recognizable on the sagittal and horizontal sections, were also localized on a number of frontal sections and their distance was measured along these. Being sections parallel to the base of the idealized cone, the frontal sections appear as arcs of circles on the flattened-out conical surface be-

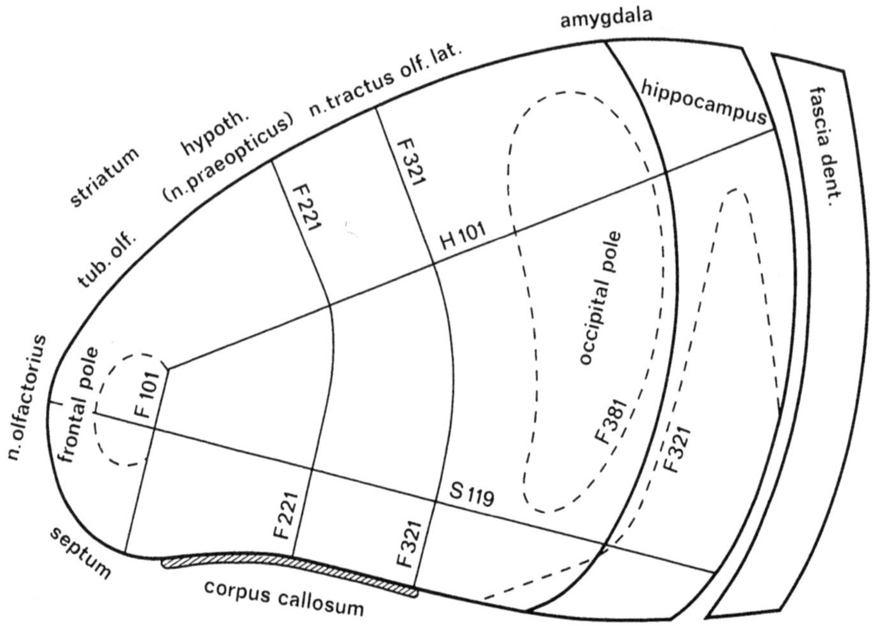

Fig. 71. Planar map of the mouse cortex, right hemisphere. The map was designed in such a way as to make the distortion due to the flattening tolerable in most places. F101, F221, F321, S109 and H101 are frontal, sagittal and horizontal sections of the Sidman Angevine and Pierce atlas (1971), along which distances are correctly represented on our map. Other sections (F381, F321) are represented on the map as closed or strongly curved lines *(dashed)*, due to bulges (F381) or folds (F321) of the cortex. The shape of the flattened cortex is roughly triangular. The orientation is: medial - *down*, lateral - *up*, frontal - *left*, and occipital - *right*. The regions of the brain bordering the cortex are also indicated, see Chapter 2

tween the principal sagittal and horizontal sections. Only in the region of the map representing the flat interhemispheric cortex, as well as in that representing the basal cortex, are the frontal sections drawn as straight lines.

The dotted lines represent singularities in the cortical curvature, where frontal sections appear either as closed lines (frontal and occipital pole) or as curiously bent lines (in the hippocampus).

The map should be read as follows. Along the selected sagittal, horizontal and frontal sections, drawn in the map of Fig. 71, distances are correctly represented. This is also true, with good approximation, for the whole cortex between the frontal and occipital pole regions, set off by dotted lines, including the interhemispheric and basal cortex. Therefore areas in this region, too, are almost correctly represented. Also, the width of the hippocampus being quite constant (measured along the sort of cut that hippocampal physiologists call a hippocampal slice), and its length being quite correctly represented on the map, the long strip of hippocampal cortex on our map may be considered a good representation of that region. The dentate region is equally well in proportion, but it is nowhere continuous with the cortex, as we have already remarked. It should be remembered that it is not even oriented in the plane of the cortex, but at right angles to it.

The error of the approximation varies from place to place and can be estimated as follows. Within a region of the map outlined by lines all representing distances correctly (such as the region between the lines marked F 221, H 101, F 321 and S 119) the real distances are slightly larger than the corresponding distances on the map, because of a slight bulge of the cortex in this region which has been neglected when the cortex was idealized as a cone. The difference can be computed from the radius of curvature of the bulge that has been neglected. For a segment of 1 mm length on the cortex it lies below 5% for most of the central region of the map. Only in the pole region may it approach or even exceed 20%.

The mapping of the serial sections outside the standard sections (e.g. in Fig. 73) was performed by an obvious process of interpolation.

Thickness of the Cortex. The thickness is defined as the distance of the surface of the cortex from the level of transition cortical grey-subcortical white, measured along a line perpendicular to the surface of the cortex.

There are various difficulties involved in this measurement. In some regions, such as the cingulate cortex or parts of the hippocampus, a line

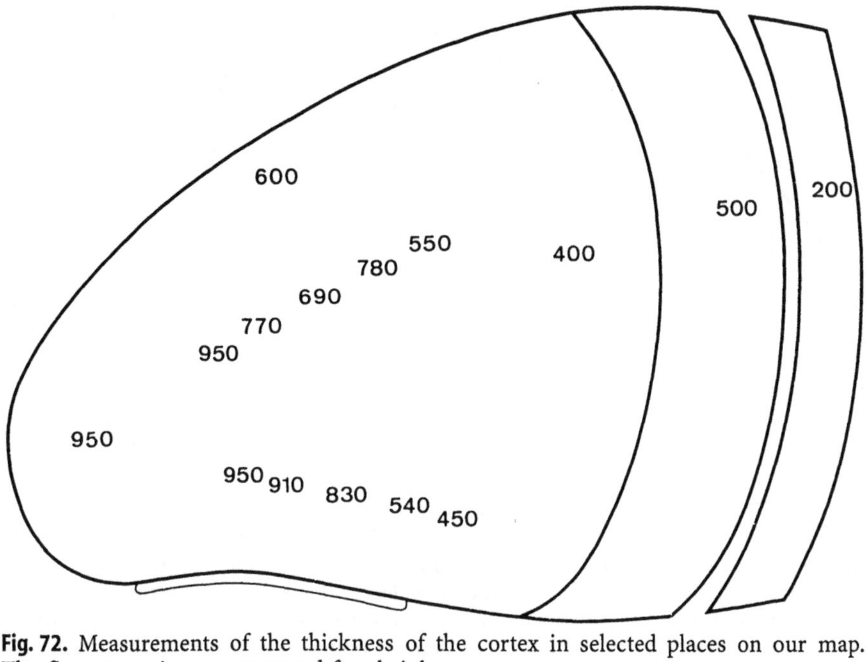

Fig. 72. Measurements of the thickness of the cortex in selected places on our map. The figures are in μm, corrected for shrinkage

perpendicular to the surface does not correspond to what one would intuitively like to define as the vertical direction of the cortical organization, namely, the direction of the descending axons of pyramidal cells. This is a serious discrepancy only in the region of the cingulate cortex, which we have therefore avoided in our sampling. Another difficulty may arise in parts of the hippocampus, where the surface of the cortex can be recognized only microscopically because of its intimate adhesion to the surface of the fascia dentata. Finally, unless histological sections are specially prepared from blocks cut perpendicularly to the surface, most sections do not show the true thickness of the cortex anywhere and some sections only in some places, while the majority of measurements has to be corrected for the angle of the section with respect to the vertical direction in the cortex. This can be determined by comparison of selected places as they appear on the three standard serial sections. Again, for the determination of this angle as for the construction of the cortical map, we used the SAP atlas.

Figure 72 displays some measurements of the cortical thickness on the cortical map. Generally, the highest values are obtained in the frontal regions and the lowest on the occipital pole and in the basal regions. By di-

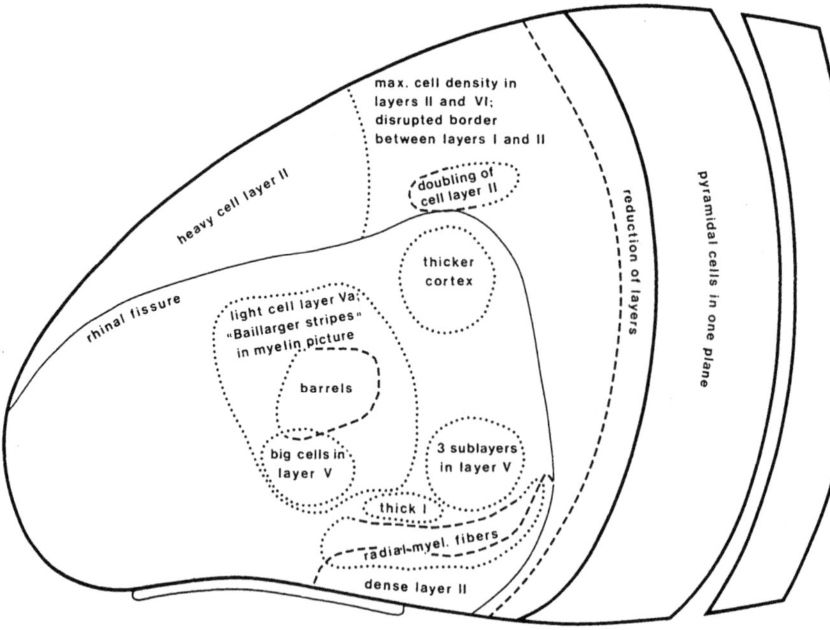

Fig. 73. Some local variations of cortical layering which readily meet the eye. *Heavy lines* outline of the cortex, hippocampus and fascia dentata. *Thin continuous line* the rhinal fissure. *Dashed* well-defined, sharp boundaries; *dotted* outlines of regions with characteristic layering, but without well-defined boundaries. Some of the regions on this map can readily be identified with the areas of Caviness' (1975) map. (see Fig. 74)

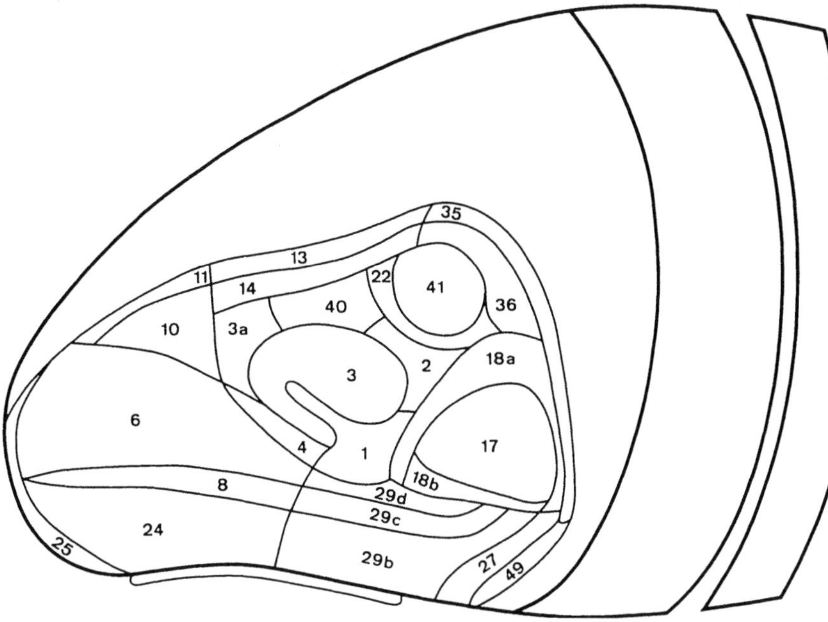

Fig. 74. The architectonic map of the mouse cortex by Caviness (1975) redrawn on our planar map. The numbering of the areas is based on the one proposed by Brodmann (1909) for the human cortex. The correspondence between the areas in the mouse and those in larger brains was established on the basis of known physiological properties as well as on the pattern of their projections

viding the cortex into sectors of roughly equal areas and taking the average thickness for each of the sectors, the average of the whole cortex, except for the hippocampus and fascia dentata, can be computed as 0.65 mm (on paraffin embedded material). Paraffin sections are subjected to greater shrinkage than the material of our Golgi method (see Chap. 3 on shrinkage). The unfixed mouse cortex should have an average thickness between 0.8 and 0.9 mm.

31.1 The Architectonic Map

Figure 73 displays on the cortical map the major distinctions we have described.

Clearly, neither we nor our predecessors have been entirely misled by accidental variations of cortical structure nor by perceptual effects, as some radical critics of "architectonic" studies have surmised ("little more than the whim of the individual student" said Lashley and Clark in 1946). Many local traits of the cortex which impressed us we found mentioned in the papers by Isenschmid (1911), Drooglever Fortuyn (1914),

Rose (1929) and Lorente de Nó (1938). The areas described and numbered by Caviness (1975), when transferred onto our planar map from a similar projection in the original paper, correspond well enough in position to most of our locally distinct regions (Fig. 74).

Chapter 32

How Cortex-Like Is the Hippocampus?

The quantitative picture of cortical anatomy which we have sketched in Chapters 4 to 23 was based on measurements and observations limited to the dorsal hemisphere, i.e. to the part of the cortex roughly coinciding with the so-called isocortex. We have made no attempt at including the allocortex, although if we had done so, the general conclusions would have certainly remained the same and the quantitative statements would probably have changed very little. In particular, we did not try to stylize the general picture of the cortex in such a way as to include the hippocampus, its most strikingly aberrant part, although we firmly believe that the hippocampus shares with the cortex not only the continuity of the network but also its most prominent features. The reason for our initial exclusion of the hippocampus is a sociological one rather than a matter of principle. The community of workers in this field has but little overlap with the one concerned with the cortex proper, and the connections to physiological questions lead to entirely different contexts in the two groups, to experimental animal psychology in one case and to cognitive studies in humans on the other. We did get involved in the hippocampus, however, when we were drawn accidentally into a meeting on the hippocampus, and the result was a paper (Braitenberg and Schüz, 1983) which we summarize briefly here. Our observations on the hippocampus are sporadic, not related to main stream thinking in this field and were not followed up by the research which they perhaps deserve. They do fit, however, into our general picture of the cortex.

The continuity of the plane of the cortex into that of the hippocampal formation can be clearly seen in Figs. 2 and 75, which also show how the convexity of the cortex turns into a concavity in the hippocampus. The most striking characteristics of the hippocampus is the alignment of most of the cellular elements in a narrow layer, the layer of pyramidal cells which appears as a dark band in the Nissl picture. By contrast, the cells are scattered throughout the layers (except for the first) in the ordinary type of cortex. This is all the more remarkable, since most of the neurons are very similar in the two types of cortex, being rightly called pyramidal in both cases (Fig. 76). However, hippocampal pyramidal cells, when they are located in the CA3 region, i.e. close to the fascia dentata, are distinguished from all other pyramidal cells by peculiar spine-like excrescences on the lower part of their apical dendrite. These are larger than ordinary spines, and more lobated, but serve the same purpose as

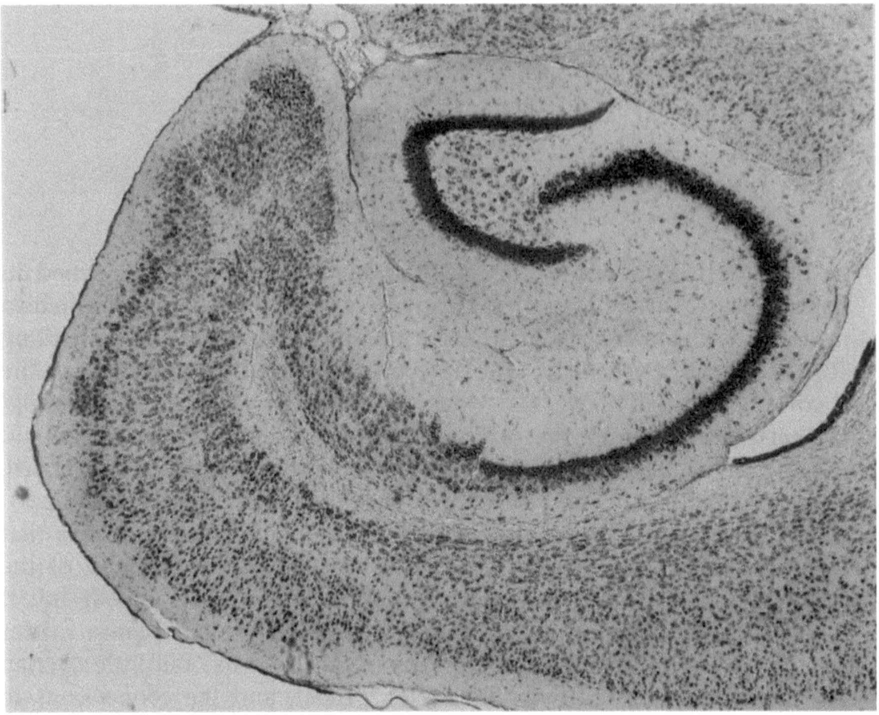

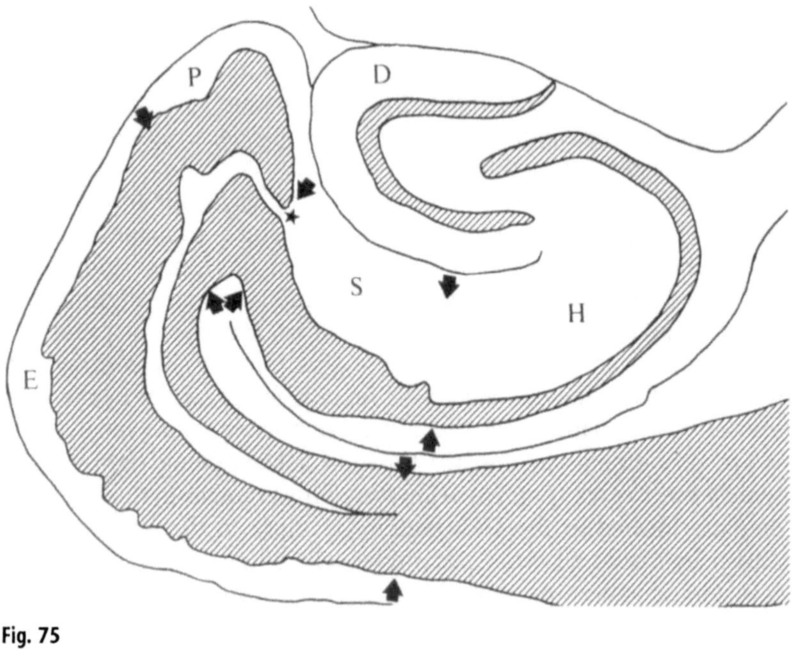

Fig. 75

carriers of synapses, only these hippocampal super-spines specialize as recipients of equally gigantic axonal swellings, those of the "mossy fibres" which actually gained their name from these "moss-like" appendages (Figs. 76, 77).

The mossy fibres, which are axons of "granular cells" in the fascia dentata, together with the fibres of the "perforant path" which are afferent to the granular cells, and the "Schaffer collaterals" emanating from the CA3 pyramidal cells, form a *three-link chain* which has a unique distinction in the cortex, namely that of being arranged in one direction only. Nowhere in the cortex have we seen such a striking exception to the prevailing principle of a homogeneous distribution of fibres in all the directions of the cortical plane. Two of the links of the three-link system, mossy fibres and Schaffer collaterals, lead from the margin of the cortex, formed by the hippocampus and crowned by the fascia dentata, towards the main part of the cortex, while the third link, the perforant path, seems to derive signals from the hippocampal-cortical stream and put them back into the fascia dentata, where it originates. This looping arrangement and the feedback (most likely positive) which it implies has inspired theorists, who put it in relation with the special role of the hippocampus in memory (suggested by some clinical observations and, to a lesser extent, by animal experiments). The parallel arrangement of many loops of the three-link system, each in a plane perpendicular to the long axis of the hippocampus, has also suggested a peculiar preparation to electrophysiologists, the "hippocampal slice". What they mean by this is a flat excision containing a sample of the three-link system, oriented in such a way as not to sever the loops. In such slices, the action of one element of the loop on the next following one can be observed directly, and even changes of the strength of this influence can be induced by imposing something like a primitive "learning". (Bliss and Gardner-Medwin, 1973; Bliss and Lømo, 1973; Schwartzkroin and Wester, 1975; Bonhoeffer et al., 1989, and many others).

We were interested in the unidirectional arrangement of neurons in the hippocampus and wanted to assess how many of the hippocampal neurons were involved in it. For this purpose we examined tangential sections through the upper and the lower tiers of the hippocampal cortex. It turned out that the network of axon collaterals of the hippocampal py-

Fig. 75. Continuation of the cortex into the hippocampus. *Above* Nissl preparation; *below* schematic chart of the same. It can be seen that of the two main layers of the entorhinal cortex *(E)* the more superficial one ends *(asterisk)* in the region of the presubiculum *(P)*, while the lower one turns into the subiculum *(S)* whose cell layer then flattens to form the dense band of pyramidal cells in the hippocampus *(H)*. The *arrows* indicate the borders between these regions. *D* is the dentate gyrus. (Braitenberg and Schüz, 1983)

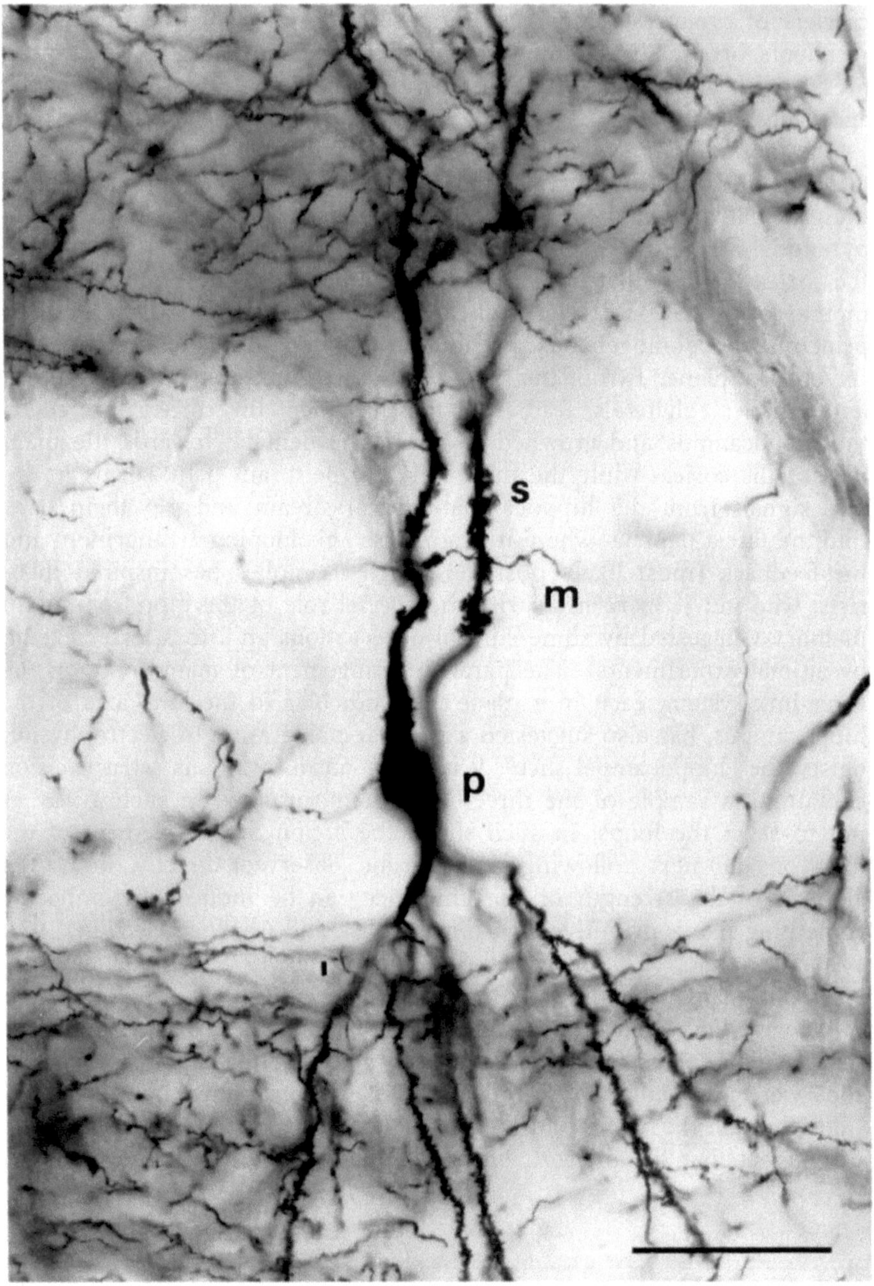

Fig. 76

Fig. 76. Pyramidal cell in the CA3 region of the mouse hippocampus, Golgi impregnation. Many fibres in the stratum moleculare *(above)* and stratum oriens *(below)* are also stained. The layer in between contains the mossy fibres, axons of cells in the dentate gyrus. One of them is stained *(m)* and shows a piece of the "moss" (= giant synaptic swelling of the axon) making contact with the apical dendrite of the pyramidal cell *(p)*. At this level, there are peculiarly large and ramified spines *(s)* on the apical dendrites of hippocampal pyramidal cells. Bar: 50 μm

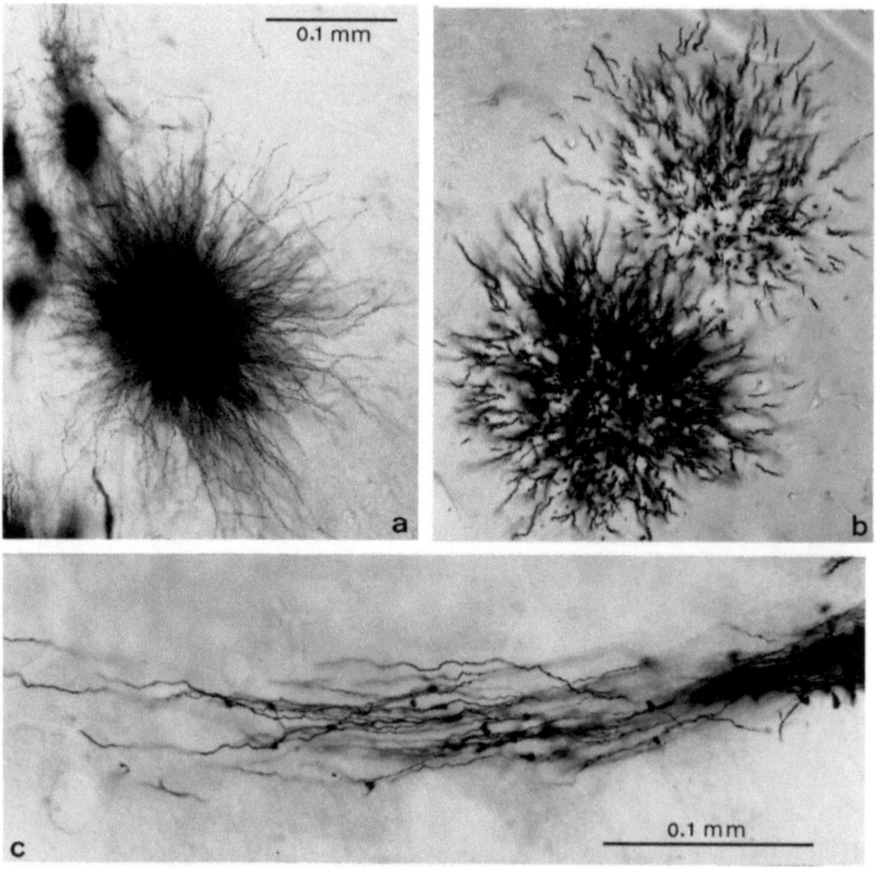

Fig. 77. **a.** "Golgiograph" showing axons emanating from a nucleus of precipitation in a Golgi preparation. Tangential section through the stratum oriens of the hippocampus. The axons do not seem to be preferentially oriented, either in the long axis of the hippocampus or in the direction of the mossy fibres. **b.** Two Golgiographs showing apical dendrites emanating from two lumps of precipitation in a Golgi preparation. Tangential section through the molecular layer of the hippocampus. Again, there is no preferential orientation. **c.** A bundle of mossy fibres above the layer of pyramidal cell bodies, CA3 region of the mouse hippocampus, Golgi stain. These fibres all run in the same direction, across the long axis of the hippocampus (in the plane of Fig. 75). (Braitenberg and Schüz 1983). See also Fig. 76

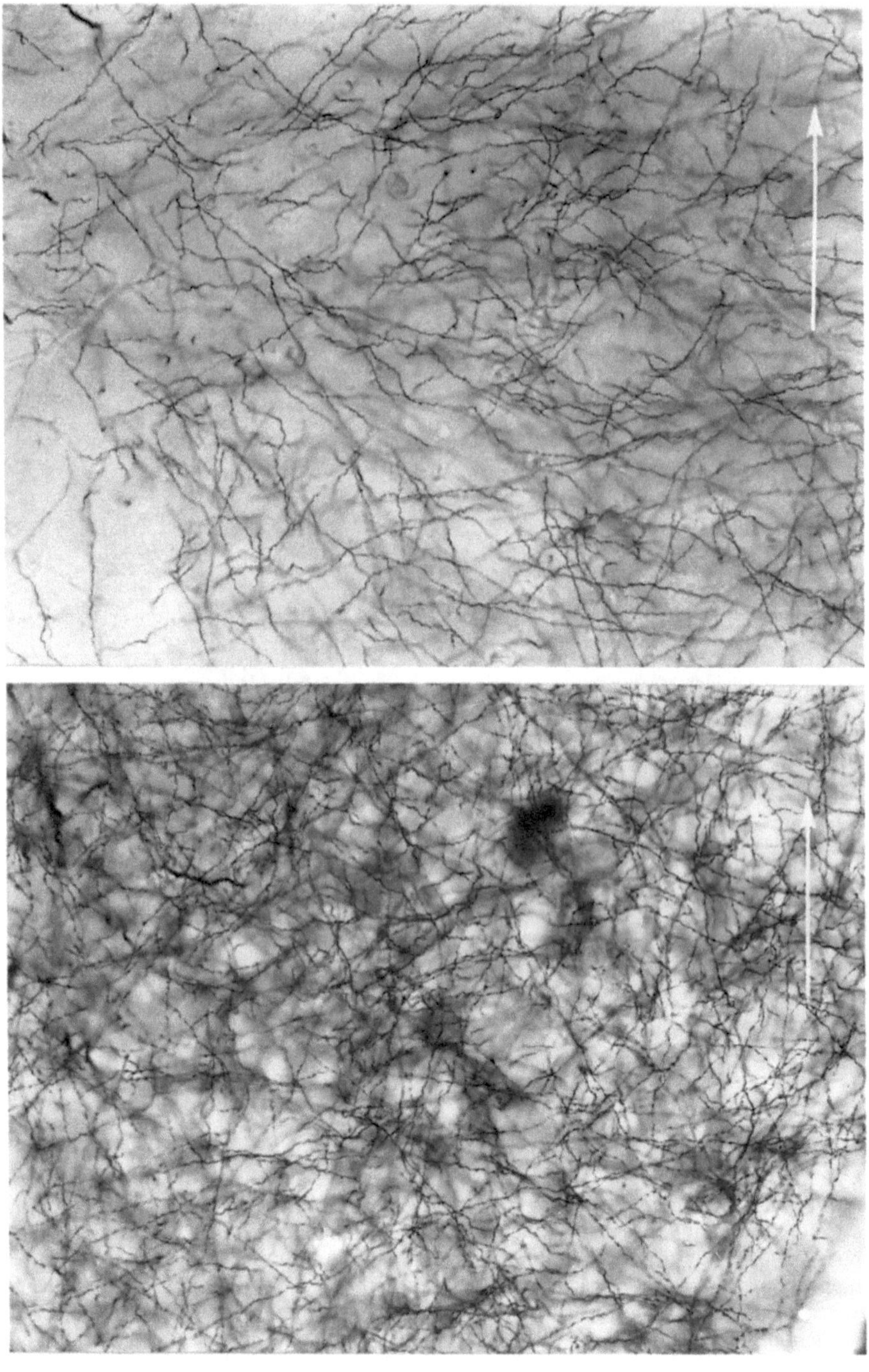

Fig. 78

ramidal cells is no more oriented than that formed by similar collaterals in the rest of the cortex (Fig. 78). Also, examining so-called golgiographs (Fig. 77), we could not detect any anisotropy. By this term we mean the lumps of silver chromate precipitation found occasionally in Golgi preparations, which locally induce the staining of all the axons or all the dendrites (or both) passing through. Thus we concluded that it is only a small proportion of the fibre population in the hippocampus which is oriented in the direction perpendicular to the long axis of the hippocampus, like the fibres of the three-link system perforant path-mossy fibres-Schaffer collaterals. We also gained the impression that the hippocampal slice is a brutal excision, one that cuts the large majority of the fibres, with the advantage, possibly, of isolating one system of connections which might otherwise be submerged among the others.

More precisely, we estimated that the synapses in the mossy fibre system are less than 1% of all the synapses in the hippocampus. The majority of the synapses are, most likely, synapses between pyramidal cells like in the rest of the cortex, with the great divergence and convergence which we discussed in Chapter 20. Most of these synapses reside on small spines of the common kind. The number of synapses per neuron is higher in the hippocampus than in the rest of the cortex: comparing the density of synapses with that of neurons we obtained 18000 synapses per neuron, more than twice the figure for the isocortex.

The number of pyramidal cells in the mouse hippocampus (calculated from the density of the cell bodies in the pyramidal cell layer and from the surface area and the thickness of the same) is 3×10^5. The divergence between dentate gyrus and hippocampus in the mossy fibre system is not great: each granular cell contacts about 50 CA3 pyramidal cells. Conversely, about 200 granular cells converge onto one CA3 pyramidal cell.

Fig. 78. Tangential sections through different levels of the mouse hippocampus, CA3 region: stratum radiatum *(above)*, stratum oriens *(below)*. Golgi stain. The fibres run in all directions. The arrows indicate the direction of the mossy fibre – Schaffer collaterals pathway. The *length of the arrow* corresponds to 0.1 mm in the preparation. (Braitenberg and Schüz, 1983)

Chapter 33

Summary of Statistical Anatomy and Conclusions Thereof

From this point on we shall not present any new data but will try to make theoretical use of the facts we have assembled. It will be helpful to first review some ideas scattered in different chapters in the descriptive part of our book, or implied by the data there.

a) The *number of neurons* in the mouse cortex exceeds the number of input elements by at least one order of magnitude. In larger brains the number of sensory input fibers is no more than one hundredth, in the human brain probably less than one thousandth of the number of cortical neurons. One gains the impression that the information capacity of the cortex is not related in any simple way to the capacity of the sensory channels (such as might be calculated from the product of the range and the resolution of the sensory spaces represented), nor to any peripheral transformation of the sensory input (which one would not expect to differ so much in different animal species). Quite naturally, we are led to the idea of the size of the cortex reflecting the number of things and situations, or more generally, the number of different concepts an animal is in a position to deal with. There are some puzzles in comparative anatomy, though, complicated by scant knowledge of comparative behaviour: is the conceptual world of an elephant really so much larger than that of a mouse as to justify a brain six thousand times bigger? Are we humans really inferior to elephants and whales, with our brains (and presumably cortices) weighing about one quarter as much as theirs? Or could it be that in reasonably sized animals (including man) the evolutionary pressure leading to bigger brains is counter-balanced by a pressure against big skulls, which are vulnerable and mechanically cumbersome, whereas in the bulkiest mammals brains may expand unchecked, perhaps at the expense of internal finesse?

b) In a certain sense, the vast majority of cortical neurons are *interneurons*; not directly under the influence of sensory input and not directly involved in the motor output. Even where the input enters prominently, in the fourth cortical layer of sensory regions, at least four fifths of the synapses which the cortical neurons receive do not have an input fibre but presumably another cortical neuron as their presynaptic element (Peters and Feldman, 1976; White and Rock, 1979, 1980). In other regions we may assume that almost all synapses have a cortical neuron on both the presynaptic and the postsynaptic side. Whatever signal reaches the cortex and is relayed to the motor output from there has to pass through a very large network of interconnected neurons. The functional state of this network at any given moment determines the output to a greater extent than the input does, and even the extent to which the sensory input is at all "perceived" by the network, i.e. is able to perturb its dynamic state, depends largely on this state itself. This is all quite plausible in terms of psychology, where the interference of "cognition" with "perception" and the importance of "context" or "set" in many different kinds of stimulus-reaction experiments is well known.

c) Not only does the numerical prevalence of interneurons over the neurons directly concerned with input and output call into question the simple input-output scheme which is still implicit in some writings today, the global layout of the cortex also provides a further argument against simple, serial processing. The cortical architecture is that of a three-dimensional network in which only one direction has a special status, namely that along which the cortical layers are displayed in succession. If we were to guess which direction of the cortex corresponds to the transformation of input into output, our first choice would be the vertical direction, and we would be led to interpret the layers as being responsible for successive stages in this transformation. And yet the input areas (if we want to use this term for the so-called primary sensory areas) are arranged *in parallel* with the output areas from which signals are relayed to the motor executing organs, and the flow of information from the sensory to the motor areas, if there is such a thing, traverses the cortex in a direction of the cortical plane which is not at all distinguished by any special "wiring".

In view of this it seems more reasonable to talk of the motor output as something determined by the dynamic state of the whole cortex, and of the various sensory inputs as devices through which this dynamic state is continually updated. Motor and sensory areas have very similar layering and share the main features of their internal connectivity with the other areas of the cortex. The meaning of this we can only hope to grasp if we venture into a more global theory of the function of the cortex.

d) Having thus expressed a preference for a global view of the cortex, we have also opted for pyramidal cells as its basic structural constituents. In fact, it is mainly their long axons which tie the cortical network together. Every small region of the cortex is directly connected to many other regions, both near and distant, via the local and the long-range branches of the pyramidal cell axons. By contrast, all the other neural types have axonal arbors which mediate only narrow range interactions. We propose to consider the system of the pyramidal cells as the *skeleton cortex* and intend to relate it to the fundamental cortical operation. The other cell types interspersed among them would then be responsible for local exceptions or additions to the fundamental operation. The fact that pyramidal cells are the great majority of cortical neurons, and contacts between pyramidal cells the majority of synapses in the cortex, makes this interpretation even more appealing.

e) Synapses between pyramidal cells, which are the majority of all synapses in the cortex, reside on dendritic spines and are excitatory. We have shown that spines may be related to learning, in the sense that changes in their shape (or in their internal structure) may reflect the past history of the signals that reached them and may in turn affect their capacity to transmit signals. This leads to the supposition that *modifiable excitatory synapses* embody the essence of the cortical function. The idea that learning is the main task of the cerebral cortex is not new and is strongly supported by the role of the cortex in human language as well as by a host of observations on animals. What is new in our interpretation is the claim that modifiability is an exclusive prerogative of excitatory synapses in the cortex. The implication is that learning affects the tendency of cortical neurons to congregate into groups of elements which excite each other. If we suppose that the learning rule is such that coincident activity of neurons strengthens their synaptic coupling, there will be a tendency for cortical neurons to congregate into groups when the elementary events which they signify singly (or: which each of them represents) somehow belong together in the external world. Such groups have been called cell assemblies by Hebb (1949).

f) According to widespread opinion today, synapses are not established but are only modified by the learning process. The observations on the guinea pig, where nearly all synapses are there before the learning can possibly start, i.e. before birth, make this very likely. If the elementary learning process is the modification of the strength of a synapse, the presence of the synapses is a necessary condition for learning, and the selection of established synapses among the near infinity of possible synapses between so many neurons must strongly determine what can be learned by the brain. The 10^{11} synapses in the mouse cortex fall short by a factor 1 000 of the number that would suffice to connect each of the 10^7 neurons with each of the others. Most likely the limitation stems from the impossibility of any one cell to carry more than a certain number of synapses and perhaps even more from the impossibility of creating a network of fibres sufficient for the full set of connections.

If synapses are there initially in order to let the neurons know of each other's correlated activity, and eventually to be strengthened as a consequence of it, we should expect every neuron in the skeleton cortex to contact as many other neurons as possible. Given the numerical limitations, this would give the learning network the greatest possible chance of learning unforeseeable combinations of activity. Our calculations of the probability of one or more than one synapse between two cortical neurons indicate that multiple synapses between two given neurons are indeed rare, and that therefore the greatest possible *divergence* (and *convergence*) of signals is nearly achieved in the network of pyramidal cells. More than any other part of the brain, the cerebral cortex appears as a device for the mixing and reshuffling of patterns of activity in its interior.

g) The great divergence and convergence is at the expense of the strength of the individual connection. If every neuron receives thousands of synapses from different sources, the influence of any single afferent cannot be very strong. We do not know how many synapses must be activated on the dendritic tree of a neuron in order to meet its threshold for an action potential (estimates of this number are given in Abeles, 1982), but a single synapse is surely not enough. Evidence from multiple unit recording (e.g. in Abeles, 1982) also shows that the excitatory influences of individual neurons onto each other are quite weak.

As a consequence of this, the condition for a cortical neuron to be activated must be synchronous activity of many of its afferents. Cortical activity cannot be transmitted through chains of individual neurons connected in a serial fashion, but must necessarily involve the activation of fairly large *groups of neurons*. Just how large, is a problem also addressed by Abeles (1982). It involves the open question of the thresholds of cortical neurons.

The great divergence and weakness of connections within the skeleton cortex is not compatible with the idea of circuitry of the electronic kind.

h) The looseness of the connections within the skeleton cortex can also be seen in a geometrical way. From the point of view of their distribution in space, the synapses for which a neuron is postsynaptic form a *cloud* of a certain shape corresponding to the branching pattern of its dendritic tree. The synapses for which a neuron is presynaptic are also distributed in a cloud corresponding to its axonal ramification. Simplifying a little, we may suppose that within each cloud the localization of individual synapses is random. The density of synapses belonging to a single cloud is much lower than the overall density of synapses in the tissue: the synapses on the dendritic tree of one neuron are only about one thousandth of all the synapses there, and for the axonal clouds the relative density is even lower. Thus the dendritic clouds of different neurons strongly overlap with each other, and, especially in the case of pyramidal cells which have widely varying sizes, the dendritic cloud of one may be completely contained within the dendritic cloud of another. The axonal clouds permeate each other even more intimately. More important still, the axonal cloud of one neuron overlaps with the dendritic clouds of many other neurons, and conversely. Synapses are thereby established, and the probability of a synaptic connection between two neurons can be calculated from the overlap of the axonal cloud of one with the dendritic cloud of the other and (assuming their independence) from the product of their respective densities (Braitenberg and Lauria, 1960).

It is very difficult to draw, or even to envisage, such clouds of synapses permeating each other thousandfold. Therefore the illustrations of Golgi-stained neurons nicely separated in space and touching each other gently with a few synaptic endings will stay with us for some time. Unfortunately, they are apt to suggest neuronal networks which are very far from the truth (Fig. 79).

An attempt at deriving cortical connectivity in a probabilistic way from the known shapes of neuronal ramifications (Krone et al., 1986) produced functional models of receptive fields in the visual cortex which were surprisingly realistic.

i) It is not easy to provide an interpretation for the long range cortico-cortical connections which would fit smoothly into the picture of the skeleton cortex as we have sketched it. It is not clear whether the long axons of pyramidal cells which traverse the white substance are just one extreme of a distribution in which fibres of all lengths occur, or whether their role is radically different from that of the local branches of pyramidal cell axons which stay in the cortex. One thing is certain, their distribution is far from random. Apart from the strict topographical rule that governs the callosal connections, also the fibres which stay within one hemisphere seem to follow a plan which assigns to them different destinations depending on their place of origin. While within one cortical area we may describe the pyramidal-cell-to-pyramidal-cell connections in statistical terms, the connections which tie different areas together are quite specific. While the local connections between pyramidal cells depend on their distance, in the sense that the probability of such a connection is inversely related to the distance between the cells, no such simple dependence on the distance is valid in the long range cortico-cortical system. For this reason we (Palm and Braitenberg, 1979) proposed the terms *metric* and *ametric* system, or metrically dependent vs. non-metrically dependent system for the two kinds of connections between pyramidal cells, hoping to catch the essence of the distinction. Another set of terms we used is A-system vs. B-system (Braitenberg, 1978 b), to remind us that the long-range connections terminate preferentially on the *A*pical dendrites of pyramidal cells, and the local collaterals on the *B*asal dendrites. The two sets of fibres within the skeleton cortex correspond, at least in a statistical way, to the two distinct sections of the pyramidal cell dendrites.

j) The two sets of fibres in the A- and B-, ametric and metric system, if they are really distinct, suggest a two-tiered hierarchy in the organization of the cortex. This idea receives further support both from anatomy and physiology, which provide evidence for distinct *areas* in the cortex, defined by structural traits and in many cases by special relations to input and output systems. Roughly, we may surmise that information handling within an area is related to the internal connections in the B-system, while the projection of activity patterns from area to area relies on the A-

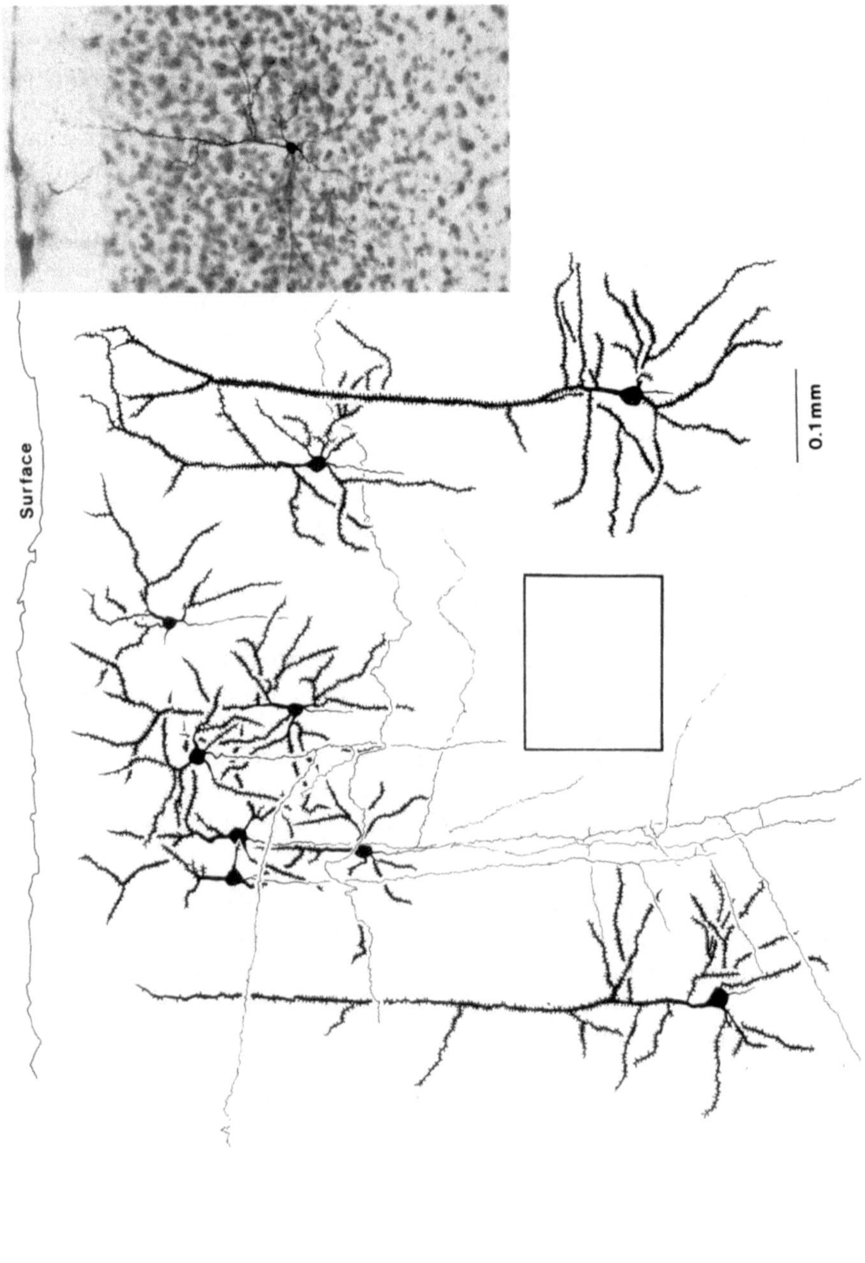

Fig. 79. Pyramidal cells from one Golgi preparation in their correct relative position. Their location in depth varies between layer II and layer VI. In some of them the descending axons and the axon collaterals are also stained. Camera lucida tracing. The microphotograph on the *right* shows at the same magnification one Golgi-stained pyramidal cell amongst a multitude of cell bodies counterstained with a Nissl stain. This is intended as a reminder of the real density of neurons in the cortex of which the Golgi picture on the *left* gives a very wrong impression. The rectangle at the *centre* shows the size of the region illustrated at high magnification, and with different staining methods, in Figs. 17 and 20

system. Since areas in many cases are organized in a somatotopic, visuotopic etc. way, it is reasonable that the fibre systems within should be "metrically dependent". The topographic representation of a sensory (or motor) space in the cortex preserves the neighbourhood relations and thus allows the cortical machinery to make use of the redundancy that resides in the structure of space itself (a moving object produces a signal which cannot but move continuously through an orderly map of a sensory space; the images of isolated objects are enhanced by some combination of lateral excitation and lateral inhibition etc.). Thus within any such cortical representation of a sensory field we expect a homogeneous, short-range fibre population, and we are not surprised to find a slightly different statistics of the intracortical fibres in the areas concerned with different kinds of sensory input (visual, auditory, somatosensory), or with the same input in different *contexts* (e.g. acoustic input in the ordinary context and in the context of language). Besides differences in the B-system, consisting of the axon collaterals of pyramidal cells, differences in the distribution of non-pyramidal cells may also play a role in areal variation. But all of this remains speculative and cannot be directly related to the different pictures which the "architectonic" analysis of the cortex reveals.

The situation is different in the A-system, in the fibres connecting different areas which may be close to each other or very far apart. These fibres mediate information between cortical representations referring to different aspects of reality, and there is no a priori reason why the topographic map of the cortex as a whole should reflect in any way an abstract relation between these aspects. We cannot think of any reason why the visual, acoustic, somatosensory, olfactory and motor representation should be in any particular geometrical arrangement on the surface of the cortex, except perhaps when they somehow share the coordinates of external space (in the human visual, somatosensory and motor cortex the up-down coordinate is represented in roughly the same direction on the cortex). We are not surprised that there is no simple dependance on distance in the A-system of interareal connections.

k) The structure of the reasoning involved in our essay is schematically shown in Fig. 80. The quantities measured are shown on the left, the theoretical conclusions on the right and the deductions which led to them in between. To our mind, the idea which fits the cortical network most admirably is that of an *associative memory.* This is, in a way, disappointing, since many of us would have preferred to think of the cortex as a great thinking machine, performing within itself the most sophisticated neuronal computation. If the cortex is mainly memory, the act of thinking must involve it in concert with other parts of the brain, and we are now called upon to unravel this more complicated operation. We are not yet prepared for this, since what little is known about cerebral physiology re-

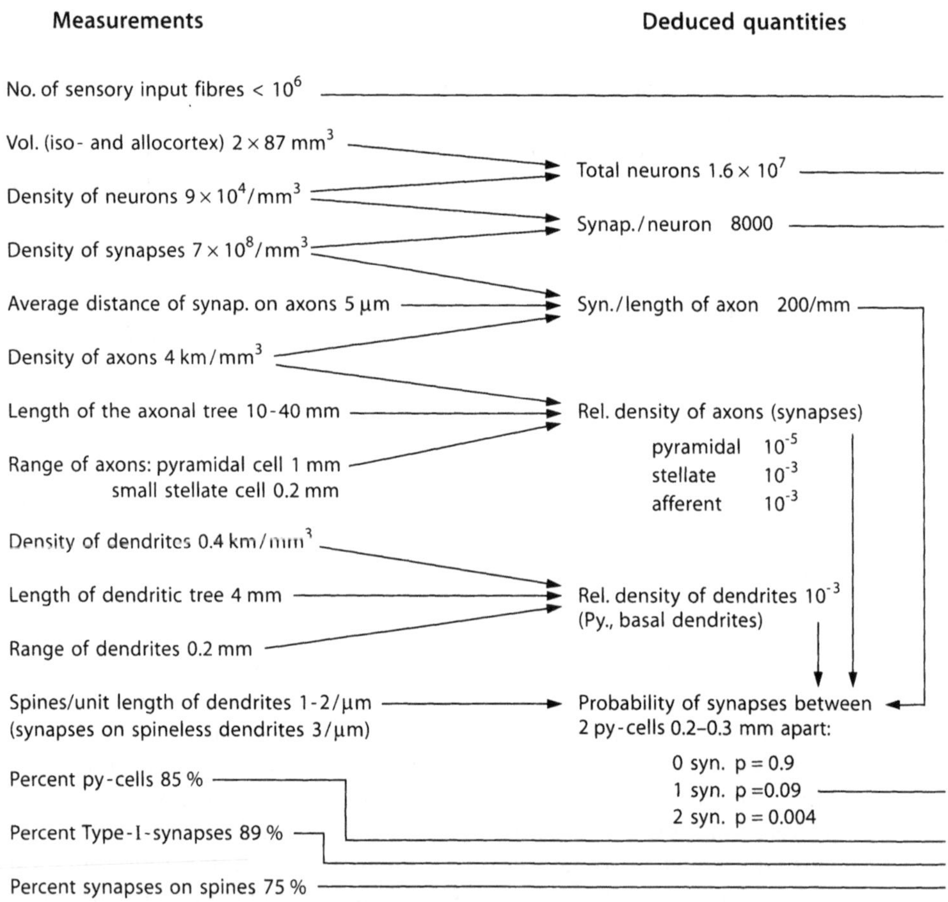

Fig. 80. A chart summarizing the measurements in the mouse cortex and the inferences made in Chapters 2 to 25. Some of the figures were rounded off within the limits of our experimental error. The exact figures are in the text

fers either to synaptic relations of a few neurons within a piece of grey substance, or to macroscopic relations of large pieces of grey substance ("nuclei", "centres") exciting or inhibiting each other. There have been a few attempts at unifying the two levels of analysis, but much remains to be done.

It is important to emphasize that the structural characteristics of the cerebral cortex summarized in Fig. 80 are not just general properties of the grey substance, but that they are quite distinctive for that level. The comparison with the *cerebellar* cortex is striking. While in the cerebrum the "skeleton cortex" of pyramidal cells impresses us with its massive positive feedback, there is not one positive reentrant loop in the cerebellar cortex: all feedback circuits there involve inhibitory synapses. Also, in

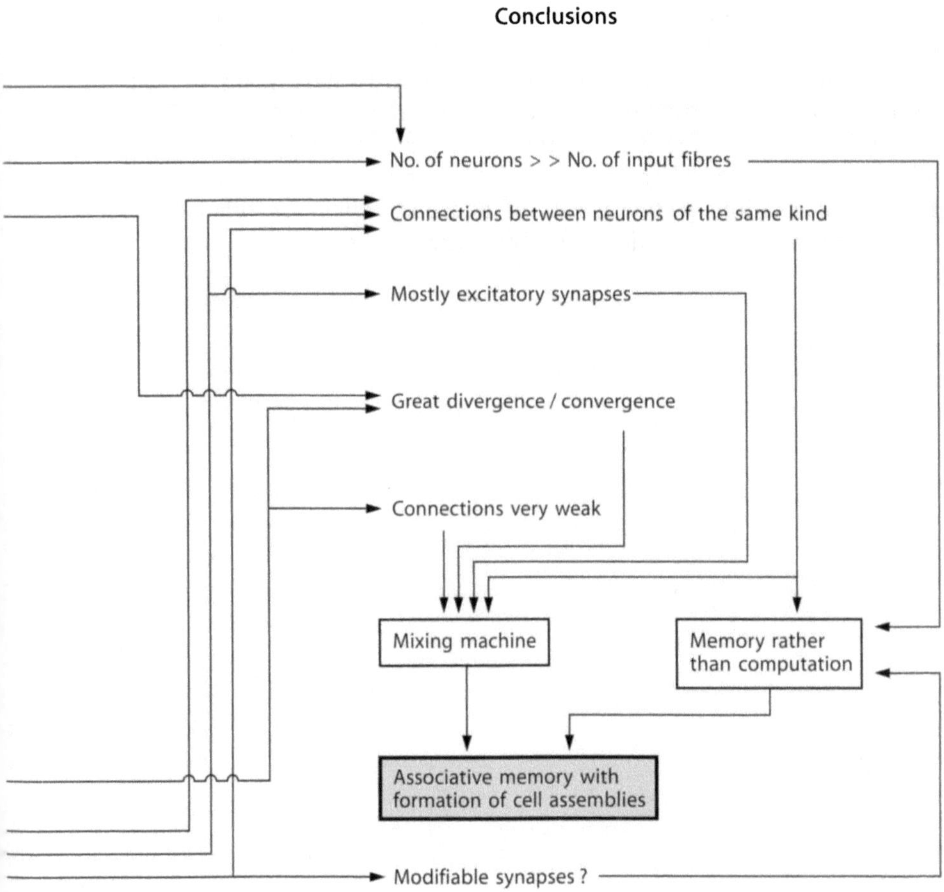

Fig. 80 (Part II)

the cerebellar cortex there are fibres of a fixed length spreading the input from any point over a constant distance (one such system of fibres excitatory, the other inhibitory, at right angles to each other). The interactions of different inputs are thus strongly dependent on their distance and relative position. We have seen that in the cerebral cortex the situation is quite different, due to the connections of widely varying range between pyramidal cells. Neighbourhood relations are largely abolished in the A-system of connections between pyramidal cells, and are relatively unimportant even in the B-system due to the probabilistic nature of the connections there. This brings us to the third striking difference between the cerebrum and the cerebellum. The abundant overlap, reciprocal permeation and even complete inclusion of one dendritic field within another is typical for the cerebral cortical network. In the cerebellar cortex, the dendritic trees of neighbouring Purkinje cells are neatly separated, each keeping out of the territory of its neighbours, with hardly any overlap. In the terminology we used in Chapter 19, the relative density of Purkinje cell dendrites is 1, while it is around 0.001 for cortical pyramidal cells, as will be remembered. There must be a meaning to such entirely different geometrical relations between sets of neurons which are otherwise very similar (e.g. both are studded with dendritic spines).

Chapter 34

Comparative Aspects: Statistical Measures in Larger Brains

The similarity between the cortices of different mammalian species has already been mentioned in the Introduction. This is why we confidently make general statements about the cortex, including the human one, based on data gained from one species, the mouse. However, we are aware of the fact that the differences in brain size between various mammalian species not only imply differences in the total number of neurons, but also in some of the other measures. Comparative aspects have already been touched upon in previous chapters. Here, we will consider in more detail how some of the measures shown in Fig. 80 vary with brain size. Unfortunately, for many of the measures, comparative data are lacking or were obtained by different authors with different histological or stereological methods making a comparison difficult. In addition, comparison of brains of different sizes is complicated by the geometric distortions which accompany the folding of larger cortices. We will not present an exhaustive comparative review here, but will content ourselves with showing some trends and invariants (Fig. 81), in as much as they play a role for the conclusions drawn in Chapter 33.

Most of what can be said about the relation between brain size and histological measures can be found in Jerison (1973). Apart from the total number of neurons, the most obvious dependency on brain size concerns the density of neurons. Larger brains have a lower density of neurons in the cortex. The exact relationship between brain size and neuronal density is somewhat controversial, but the trend towards a lower density in larger brains is unquestionable. The dependence of the neuronal density on brain weight seems to follow an inverse power function with an exponent that has been variously reported as -0.14 or $-1/3$ (Jerison, 1987).

The lower packing density is largely due to an increase in the length of neuronal processes rather than their thickness. Data by Bok (1959) show an inversely proportionate relationship between dendritic length per neuron and density of neurons. This is also suggested by a glance at electron micrographs of different species. Although some cell bodies, dendrites and axons can become very thick in large brains, the bulk of the processes in the neuropil looks very similar in small and large brains. This makes it likely that the average thickness of neuronal cell processes and therefore the total length of axons and dendrites per mm^3 is quite constant in different species. Measurements in the cat by Foh et al. (1973) which are close to ours in the mouse support this.

With increasing brain size:

Density of neurons:	↓
Length of axons and dendrites per mm³:	–
Length of axons and dendrites per neuron:	↑
Density of synapses per mm³:	–
Density of synapses along axons and dendrites:	–
Synapses per neuron:	↑
Percentage of inhibitory synapses:	–

Fig. 81. Relation between brain size and some of the measures in Fig. 80, which either increase (↑), decrease (↓), or stay (roughly) constant (–). Modified from Schüz and Demianenko (1995)

Another measure which is astonishingly constant is the density of synapses per volume. This was already evident in the comparison of different areas in the mouse (Chapter 5), but also seems to hold between species. Again, it is difficult to compare data from the literature because of different histological and stereological approaches. In spite of this, data on synaptic density in the literature do not differ by more than a factor of 2 (for reviews see Colonnier, 1981; Peters, 1987), and they differ much less – if at all – when only data from one author gained on material prepared and evaluated in the same way are compared. Cragg (1967) compared mouse and monkey and found quite similar values in both species. We ourselves had access to material from hedgehog and monkey, both prepared in the same way, and could not find any difference in the density of synapses between these two species (Schüz and Demianenko, 1995).

In view of the fact that most synapses in the cortex reside on dendritic spines, it is not surprising that the density of spines along dendrites is also almost invariant (Schüz and Demianenko, 1995).

Thus the average number of synapses per neuron is be larger in larger brains. Since the synaptic density is roughly constant, this ratio can be simply determined from the density of neurons (see also Stevens, 1989). We must therefore briefly return to the controversial relationship between brain size and neuronal density mentioned above.

According to the older literature, the density of neurons decreases by a factor of 10 from mouse to man (e.g. Tower, 1954). However, according to Rockel at al. (1980) the density of neurons decreases only by as much as cortical thickness increases, which is by about a factor of 3 between mouse and man (Blinkov and Glezer, 1968). Data by Cragg (1967) are in-between. If we take Tower's and Rockel's values as lower and upper limits, the average for the number of synapses per neuron in humans should range somewhere between 24 000 and 80 000.

Data on the monkey cortex also fall within this range. Here, both the density of neurons and of synapses has been measured by the same author (Cragg, 1967). For the motor cortex, the ratio between these values yields a number of 60000 synapses per neuron. Since the motor area is the thickest part of the cortex, the total average over the whole cortex should be lower, if we believe Rockel et al. (1980). In area 17, there are only a few thousand synapses per neuron (Cragg, 1967; O'Kusky and Colonnier, 1982), but area 17 in primates has a particularly high density of neurons and is, therefore, not representative for the rest of the cortex.

An important question is whether the increase in the number of synapses per neuron in larger brains entails an increase in the number of pre- and postsynaptic partners, or alternatively, a stronger coupling between partners. In order to answer this question, we must make some considerations on the ramification patters of pyramidal cells in larger brains. Let us take Cragg's numbers of neuronal density and assume that in the human brain, the density of neurons is about 1/6 of that in the mouse and therefore the average neuron in the human brain has 6 times more synapses than in the mouse. Such a neuron, in order to reach 6 times more neurons than a neuron in the mouse, must ramify in a volume not 6 times, but 6^2 times larger on account of the reduced neuronal density. If we assume that the neural processes spread evenly in all directions, their length would be increased by the 3rd root of 36, which is 3.3. Appropriate data are lacking, but there is evidence from Bok's data (1959) that the increase in average dendritic length per neuron goes along with an increase of the volume within which neurons ramify. In humans, the dendritic trees of large pyramidal cells can reach a diameter of 1 mm (Poljakow, 1973; Braak and Braak, 1976), four or five times larger than the largest in the mouse, and even smaller pyramidal cells in layer II occupy a region of about 0.45 mm (Buell and Coleman, 1981). Axonal trees can also be assumed to be more extended in larger brains. Intracortical axonal trees of monkeys may extend over several millimeters (DeFelipe et al., 1986). Thus, it is quite likely that the number of postsynaptic partners increases as the number of synapses per neuron.

The variation of the number of synapses per neuron (and the number of synaptic partners) with brain size supports the idea of the cortex as an associative memory. As discussed in the previous chapter (point f), a rich connectivity among neurons is necessary in order to enable the neurons to detect common activity among as many neurons as possible. If large brains were to have the same degree of connectedness as small brains, the number of synapses per neuron would have to increase as much as the number of neurons, but this would lead to impossible brain volumes (Ringo, 1991). The fact that the number of synapses per neuron does increase somewhat in larger brains may be the result of a compromise between manageable brain size and greatest possible connectedness.

Another important measure which is probably invariant is the ratio of type I and type II synapses (Schüz and Demianenko, 1995). The equilibrium between excitation and inhibition seems not to be affected by an increase in the size of the cortex.

Chapter 35

Global Activity, Cell Assemblies and Synfire Chains

The system of cortical pyramidal cells, fairly uniform in shape (and presumably in their functional characteristics), interconnected by a very large number of excitatory synapses, invites speculations as to their mode of operation. We select three ideas which have been proposed: that of a global operation, that of Hebbian cell assemblies as carriers of meaning, and that of sequential order arising in Abelesian "synfire chains". These ideas are by no means mutually exclusive and may well represent three aspects of a reality too complex to be grasped in its entirety.

35.1 Global Operation

The large number of cortico-cortical fibres, some of them spanning the length and width of the entire cortex, suggests a performance in terms of global activity patterns. With a divergence factor of the order of 10^3, as we have estimated for the mouse cortex (Chap. 20), no neuron is farther than two or three synapses away from any other neuron in the cortex, provided there is not too much overlap between the sets of neurons to which different neurons are connected (Miller, 1981; Palm, 1986). In fact, with a divergence factor of 5000, if these sets were completely disjunct, starting from any neuron one would reach $5000^2 = 2.5\times10^7$ neurons after two synaptic steps, which is more than the total number of neurons in the mouse cortex (about 10^7).

By the same reasoning, in larger, e.g. human brains, where the divergence factor may be close to 10^4 for individual neurons and the number of neurons 10^9 or 10^{10}, any two neurons may communicate with each other via no more than 2 or 3 interposed neurons.

It is doubtful whether any condition approaching such optimal global connectivity is realized in cortical anatomy, in spite of the thorough mixing and shuffling of activity patterns which occurs there, but one thing is certain. Any sufficiently large portion of the cortex is informed, either through direct fibre connections or indirectly by way of a few synapses, about the activity in the rest of the cortex. Thus there is no way of isolating pieces of the cortex from each other, and the correct description of cortical information handling would seem to be one in terms of global states of the cortex. This is important in connection with philosophical concepts ("unity of consciousness", "holographic operation", etc.) but is beyond experimental verification in present day neurophysiology, since global states can hardly be observed with sufficient precision due to the limitation of recording techniques.

Theoretically, among the infinity of possible global states of the cortex, there may be states characterized by maxima or minima of some quantity referring to the entire activity pattern, e.g. number of synchronously active neurons, or variance of the activity over the entire ensemble, or negative entropy in the sense of information theory, "energy", or any such thing. As a consequence of learning, these specially characterized states may correspond to meaningful things or events in the outside world, and the cortex may tend to fall into these states or produce sequences of them either because of its own internal dynamics or thanks to some external control mechanism. This may be taken as an explanation of perception, thought, decision and the like.

This way of thinking is, of course, inspired by certain situations known to physicists, such as modes of oscillation in vibrating bodies or patterns of magnetic domains in ferromagnetism. The theory of "spin glasses" is responsible for the greatest outburst of theoretical furore among physicists interested in brains, and has given a great boost to theoretical neuroscience (Hopfield, 1982) even if its merits do not include a detailed match with the experimental facts. Where "spin glasses" and "Hopfield nets" fall down if they are taken as models of the cortex, is in the symmetry of "up" and "down" spin, which does not find any equivalent in real neural networks. Activity and inactivity of single neurons are very different states in that activity propagates along fibres whereas inactivity does not. Also, if activity means action potential (or spike), and inactivity resting potential, clearly the time spent by a neuron in the condition of resting potential vastly exceeds the time in which it produces action potentials. Moreover, if the opposite effects of spin up and spin down in the propagation of signals through spin glass are supposed to mimic the effects of excitation and inhibition in real nerve nets, one does well to remember that in the cortex the number of excitatory synapses greatly outnumbers that of inhibitory synapses, and the networks mediating one and the other interaction have very different statistical and geometrical properties. Recent developments in spin glass type neural networks indeed take these issues into account (Amit, 1989; Golomb et al., 1990).

35.2 Cell Assemblies

When Hebb (1949) proposed the idea of "cell assemblies" as carriers of meaning within the cortex, he based his argument on two assumptions which were quite unwarranted at the time. The first was the existence, within the cortex, of a sufficient number of excitatory synapses to make it possible for a great variety of groups of cells to keep each other in a state of activity by exciting each other. The second was what was later called "Hebbian learning", the supposition that excitatory synapses are strengthened when the two neurons which they connect are frequently active in synchrony. For many years following the publication of his book (which almost coincided in time with the publication of *Information Theory* by Shannon and of *Cybernetics* by Norbert Wiener) Hebb's ideas were much more popular among engineers trying to imagine brain function than among the practicing neurophysiologists, who were reluctant to be told about the brain by a psychologist.

The situation has changed in recent years. Most of Hebb's guesses proved to be correct. The assumption that coincident activity leads to the strengthening of synapses between neurons received spectacular support in the work of Hubel and Wiesel (1965, also Wiesel and Hubel, 1965). The precise superposition of the input from both eyes on a common representation of visual space in the cortex was shown to be largely due to a learning process. While most neurons in area 17 normally receive input from exactly corresponding positions of the right and left retina, if the animal at an early age is deprived of the possibility of matching the pictures from both eyes, no such superposition occurs. Under such conditions, which can be achieved either by inducing an artificial squint or by covering the eyes in alternation, most neurons are later found to receive input from one eye only. Apparently, the two maps of the two retinas on the cortex are only roughly in register at first. The refinement is obtained through a neuronal mechanism which assigns to each cortical neuron such inputs from the two eyes as have proved to pertain to the same point of the visual field through their always being active or inactive at the same time. Here statistical correlation is directly translated into neuronal connections, just as Hebb had postulated, even if the mechanism of such early visual imprinting may not be quite the same as that which is later operative in ordinary learning. Outside the visual cortex, the most impressive demonstration of how correlations in neural activity are translated into changes in neural connectivity were obtained in the somatosensory cortex (Wang et al., 1995).

Another set of facts which proved Hebb right came from the sort of cortical histology which we propose in this book (Braitenberg, 1977, 1978 a). The very diffuse coupling of the neurons in the "skeleton cortex", the preponderance of excitatory synapses, the suspicion that the spines, carrying most of these synapses, are elements of plasticity, are all quite in accordance with the kind of nerve net that could accommodate Hebb's theory.

Briefly, we sketch the theory of cell assemblies as follows. The basic tenet is that the "things" and "events" of our experience, the units of meaning as they appear as "morphemes" in linguistics, do not correspond within the brain to individual neurons, but to groups of neurons called cell assemblies. Instead of a spike (or a burst of spikes) in a single neuron signalling the presence of a "thing" in the environment, we imagine the "ignition" of a cell assembly to serve the same purpose. By ignition we mean the rapid spread of activity to all the members of the assembly, triggered by the activity of any sufficiently large subset of them. For this to happen, we must postulate excitatory connections between the members of the assembly, and we have no difficulty in doing so once we have identified cortical pyramidal cells as the substrate in which cell assemblies develop.

The ignition of an assembly starting from any (sufficiently large) part of it is what makes the idea appealing to psychologists, who are familiar with the phenomenon of figure completion, or more generally, with the tendency in perception to perceive things of which we have previous experience, even if only partial evidence is in reality present in the sensory input. The ignition of the assembly takes the place of the spike in the single neuron, which in some theories is responsible for the all-or-nothing character of recognition and of decision.

The postulated law of "Hebbian learning", learning by association within the assembly, sees to it that assemblies form on the basis of experience. If some elementary perceptions often occur together, their internal representatives, say, single neurons, will unite into an assembly. The fact that they are often active in synchrony is evidence for their belonging to a thing or an event of the external world which has a certain inner coherence. Thus cell assemblies turn into representatives of discrete items of experience, which are likely to be relevant in behaviour because of their consistency and frequent occurrence.

Since we continue to learn and to acquire new concepts as long as we live, the number of cell assemblies should increase with age. This is indeed the case, if we interpret in this sense the steady increase of "dimensionality" in the EEG (Anokhin et al., 1996).

Cell assemblies are not necessarily confined to a narrow region of the cortex. Some may be completely contained in a single cortical area, as in the case of a cell assembly representing a certain combination of sounds which probably involves only neurons of the acoustic area. In most cases, however, a "thing" or an "event" has various aspects for which various sensory modalities are competent: a dog is a "thing" that has a certain appearance and a certain smell, barks in a certain way, feels furry to the touch and may bite: the cell assembly that means "dog" must be composed of neurons localized in different portions of the cortex. This, by the way, is one of the reasons why experimental verification of cell assemblies at the level of single-neuron neurophysiology is so difficult. Another reason is that different cell assemblies may occupy macroscopically quite the same space in the cortex if the different sets of neurons of which they are composed belong to the same neuronal population.

A cell assembly may include cortical neurons which have axons connected with the motor output organs. The "event" represented by the cell assembly would then include the motor response as well as the perception which lead to it. We may also think of such a combined motor and sensory cell assembly as representing a perception for which a motor act is essential, like touching, or scanning a visual scene with one's gaze.

This double aspect, sensory and motor, of cell assemblies throws a new light on the strange geometry of the cortex, where, as we have already noted, the motor and sensory areas are arranged in parallel rather than in series. We notice how misleading the scheme input-elaboration-output may be, implicit in the old tripartition of psychology into perception-cognition-action, and how even more restrictive our thinking is in terms of stimulus-response. We prefer to see the cortex as the microcosm in which an image of the macrocosm is set up and continuously updated by learning, the discrete elements of the representation being the cell assemblies which stand for the discrete perceptions and actions in which we code our environment and our behaviour. Most of the time perception is not followed by action (not only in humans, who may sit motionless for hours reading or watching TV), but merely incorporates information in the microcosm, updating its internal structure. Conversely, action is not always an immediate consequence of perception (especially in humans), but may take its origin from the ruminations within the microcosm, with the consequence of updating, so to speak, the macrocosm when its fit with the microcosm is not satisfactory.

If we imagine cell assemblies in the cerebral cortex in more detail, we discover a complex situation (Braitenberg, 1978 a). Different cell assemblies may *overlap*, in the sense that they may share some neurons, and still remain functionally independent: one may ignite and the other remain quiescent, if the overlap is less than the number of neurons which is required to meet the threshold for ignition. The degree of overlap is likely to be a measure of the relatedness of the concepts represented by different assemblies. There may be *inclusion* of one cell assembly in another, and they may still ignite in different situations, if the one which is included does not provide enough excitation for the ignition of the larger one. Within a cell assembly there may be several *subassemblies* of neurons which are more strongly connected than those in the main assembly. For a set of neurons to be a *cell assembly*, it is not necessary for each to be connected to all of the others. It is sufficient that each neuron excites some of the members of the assembly and is in turn excited by some of the others. This does not exclude *spurious* assemblies, which may be composed of a collection of separate true cell assemblies. For a *true* cell assembly, a further condition is that it cannot be cut in two parts without severing some of the connections between its members. A cell assembly may be *strongly* or *weakly* connected, depending on how much excitation a member of the assembly provides on average for each of the other members.

Whether a cell assembly ignites or not, depends among other things on the thresholds of the component neurons. We may consider variable thresholds – not a very plausible supposition, physiologically speaking, but one that may be approximately realized by auxiliary control mechanisms providing background excitation or inhibition. It will be seen that the ignition of cell assemblies depends on the "threshold" of the neurons and could be triggered or smothered by external *threshold control.* We shall say that a cell assembly *holds* at a certain threshold Θ, if, once ignited, it is extinguished only if the threshold is raised beyond Θ. Different assemblies may hold at different thresholds; a subassembly may hold at a higher or at a lower threshold than the parent assembly, etc.

It is appealing to imagine a threshold control which is itself controlled by the activity of the network (Fig. 82). The thresholds Θ of all the neurons are set at a level depending on the global activity A, in the simplest case by assuming proportionality between Θ and A. This will tend to keep the activity constant and will prevent the activity from spreading through the entire network when a cell assembly ignites. Clearly, the activity of a cell assembly is more likely to survive the raising of the threshold than other kinds of activity are, because of the abundance of positive feedback within the assembly and the consequent high level of excitation and synchronization of the neurons involved. Thus, automatic threshold control will make active cell assemblies stand out over the rest of the activity, or, to use a philosophical metaphor, it will discover and isolate *ideas.* In addition, there will be a strengthening ("reinforcement") of ideas represented in cell assemblies, for, as long as Hebbian learning is in operation, the strong activity within an active assembly must strengthen the synapses between the neurons of the assembly the more effectively, the more often it is activated. For a simulation study of this phenomenon see Amit and Brunel, 1997.

The idea of global threshold control depending on the average activity of the whole network (or large parts of it) is reminiscent of the way of thinking which we have already sketched under the heading Global Op-

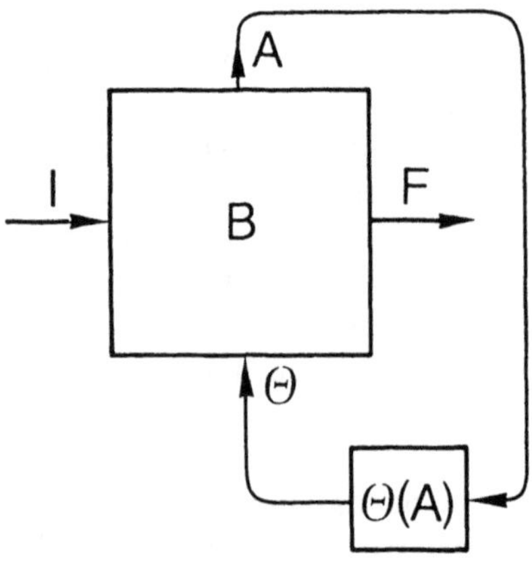

Fig. 82.
Threshold control. A brain B with input I and functional output F has in addition an output A which signals the total activity. This is fed into a mechanism of threshold control which sets the thresholds of all the components at a certain level Θ. The form of the dependence $\Theta(A)$ may vary. See text

eration. There is a difference in emphasis, however. The (presumably) explosive nature of the ignition of a cell assembly, and in particular the holding of the activity in a cell assembly once it has ignited, are spectacularly "non-linear" phenomena (as an engineer would say) which make it possible to talk about cell assemblies with little regard to the activity of the rest of the cortex. We may think of ideas incorporated in cell assemblies as having their own dynamics, their own independent evolution and, perhaps, their own rules of succession. What the threshold control does, is to keep ideas separate, preventing other cell assemblies from igniting when one is active already.

It may do more, however. The dependence of Θ on A (Fig. 82) may not be one of simple proportionality. It could be $\Theta(\mathrm{d}A/\mathrm{d}t)$, which would make it particularly sensitive to explosive ignition, or $\Theta(A, t)$, with an external programme imposing a periodic variation of the thresholds, especially when not much is happening otherwise in the cortex: this has been suggested as an explanation of some of the oscillations (in the alpha or theta band) which are apparent in the EEG (Klimesch, 1996; Kirschfeld et al., 1996).

An interesting possibility is a strategy of threshold control which works on the hysteresis in the curve relating activity to threshold (Fig. 83). If the thresholds are lowered starting from a very high value, the number of active neurons should increase progressively (we imagine a certain distribution of

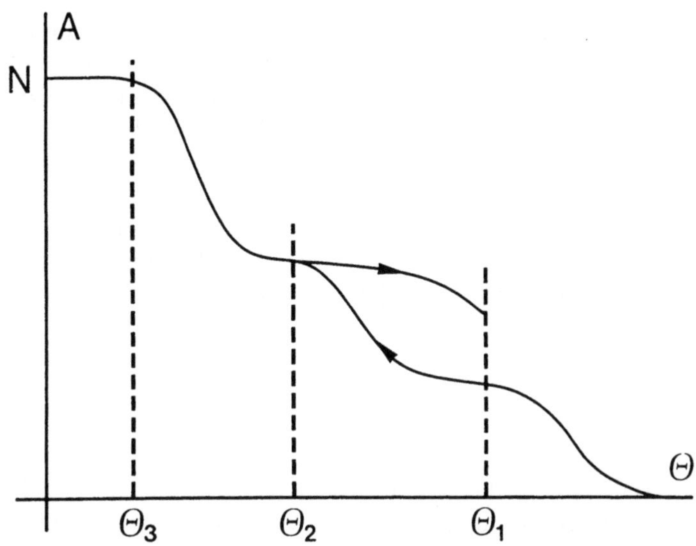

Fig. 83. Dependence of the number of active neurons A on the thresholds Θ. If the threshold is lowered first from $\Theta 1$ to $\Theta 2$ and then raised again *(arrows)*, there may be hysteresis because of self-sustained activity in the brain. N is the total number of neurons. $\Theta 3$ is the threshold at which all the neurons become active, i.e. the brain falls into epilepsy

excitation on the neurons, reflecting the internal activity of the cortex and the input). It should increase more rapidly when cell assemblies ignite. If the thresholds are then raised again, activity will diminish, but activity in cell assemblies will persist even at thresholds above the threshold at which they ignited, because of the excitation reverberating in the assembly. If the threshold is set at one of the values for which the hysteresis is maximal, this will favour active assemblies.

Cell assemblies are static entities, even if the reverberation of activity within an assembly, or the sequence in which the components ignite, may impose a certain temporal order. But the problem remains as to how a cell assembly is extinguished again once it has ignited, and how sequences of cell assemblies (of ideas in thought, of morphemes in language, of elementary motor acts in movement) are organized. We have already proposed a scheme called the "pump of thoughts" (Braitenberg, 1978a, 1984), which makes the periodic action of threshold control responsible for the chaining of ideas into trains of thought. But all of this is pure speculation, justified only by the unquestionable observations of psychology, which eventually have to be explained. We do organize our experience into discrete entities (things, concepts, ideas), we do make use of binary variables, most evident in language (we either hear a *b* or a *p*, not anything in between, we use the genitive or the dative with a certain preposition, not a mixture of the two) and this speaks in favour of discrete binary entities in the brain for which Hebbian cell assemblies are very good candidates (among other things, they serve admirably in linguistics as the bistable markers which are required in generative grammar, as we shall see in our last chapter). But we are also able to recite long poems almost automatically, to play music for hours with astonishing precision, to produce no end of speech. Something like the pump of thoughts, if it does not work the way we described it, must, on the sheer force of the evidence, be postulated in any case.

In this we are not alone. Mechanisms for the control of cell assembly dynamics have also been proposed, for instance, by Miller and Wickens (1991), Miller (1991, 1996), Wickelgren (1992), Palm (1993), Wickens et al. (1994).

35.3 Synfire Chains

The problem of serial order in behaviour (Lashley, 1951) was also felt by Abeles in his two monographs of 1982 and of 1991, where a solution is advanced in terms of cooperative activity in sets of neurons, analogous to Hebb's cell assemblies but with a temporal dimension added, and quite compatible with them.

The main idea is transmission of signals not through chains of individual neurons, each strongly exciting the next, but through sequences of *sets* of neurons, each set having sufficient, and sufficiently specific synapses on the neurons of the next set as to uniquely determine the succession. Transmission through the meagre chains of individual neurons is called the "dedicated line hypothesis", while the parallel transmission through sets of neurons received the name "synfire chain". The route taken by a visual signal from a cone through the layers of the retina and the lateral geniculate nucleus to the visual cortex is an example of a dedicated line. There may be some reciprocal influence between neighbouring lines, but lines do not cross and the character of "projection" of geometrically defined patterns is preserved on the whole, each element along the line being clearly associated with one point of the projection.

Beyond the arrival of signals in the cortex, transmission, in spite of ideas to the contrary, may be rather of the nature of synfire chains: complex intersecting routes through a network which no longer has the simple geometrical relation to the input space, and where individual neurons do not stand uniquely for certain properties of the input, but acquire their meaning only in combination.

The synfire chain hypothesis is appealing for various reasons. Like the idea of cell assemblies, it may be invoked to explain the fleeting, inconsistent relationship between many cortical neurons and the events in sensory spaces. Both hypotheses admit participation of one and the same neuron in different cell assemblies or synfire chains, as the case may be, and therefore would not predict a strict dependence of the activity of single neurons on unique input configurations. Synfire chains may explain how sequences of activity patterns (in reciting by heart, playing tennis etc.) are remarkably stable and precise, in spite of occasional failure of individual neurons, which is biologically plausible. If the sequencing is not the responsibility of isolated long chains of neurons, but of chains of events which are somewhat statistical in character, we expect a much more robust operation.

At every step in the transmission through a synfire chain, the synchronous arrival of impulses in the set of neurons which is to be activated next is decisive. This may be at the basis of our ability to detect and produce temporal patterns with great finesse, e.g. in the perception and production of speech. If the condition is synchronicity, temporal order is restored at every step and can thus be transmitted through long sequences. The combination of long duration with great precision is the most puzzling aspect of the "problem of serial order in behaviour".

Recent electrophysiological work provides good evidence for the reality of the synfire scheme. Recording the activity of neurons in the frontal cortex of awake monkeys, Abeles and Prut (1996) found that the firing times of neurons can maintain an accurancy of ±1 ms even after delays of over 400 ms. See also Aertsen, Diesmann and Gewaltig (1996).

It is interesting to speculate how many neurons are involved in each set of active neurons along a synfire chain. Abeles combines data from statistical neuroanatomy with physiological evidence on the thresholds of cortical neurons (in terms of the number of synchronous EPSPs they require) and comes up with remarkably small numbers, perhaps five synchronously active neurons at every step, although he is himself aware of the incertitude of his estimate.

Wether the concepts of synfire chains and of cell assemblies refer to two entirely different phenomena in the brain, or rather to different aspects of the same phenomenon with possible transitions between the extreme cases, is an open question. Cell assemblies remain active once ignited, while synfire chains burn out leaving no trace of their activity. However, it is well possible to imagine a synfire chain which has a cell assembly as the last link of the chain, thus representing a *sequence* of events which evokes a *concept*. We may also imagine a cell assembly which keeps itself active by letting excitation circulate along something like a synfire chain in its interior. One important difference, in theory, would be the output. In the case of synfire chains, if they are to time sequences of events, it must come from several links along the chain, while in the case of cell assemblies we prefer to think of a global, synchronous output triggered by the ignition of the entire assembly. However, intermediate cases are certainly possible, and we have already considered sequences of igniting cell assemblies prompted by external control. In a thougthful consideration of cortical dynamics, Bienenstock (1995, 1996) combines the principles of synchronicity and of sequencing by invoking the synchronization of parallel synfire waves as the binding principle, postulated by cognitive psychologists.

There is increasing support for the idea that periodic oscillations in the cortical EEG with a frequency above 30 Hz, which are elicited by certain sensory stimuli, are due to reverberating activity among the neurons of Hebbian cell assemblies (König et al., 1992; König et al., 1995; Müller et al., 1996; Pantev, 1995; Pulvermüller et al., 1996; Varela, 1995). What exactly the neuronal substrate of these periodic firing patterns is, and whether we are just observing an epiphenomenon of the essential function of tying neurons together in an assembly, or rather a more sophisticated spatio-temporal code, is not yet clear.

The application of theoretical physics to nerve nets is problematic because neurons behave very differently from particles, but some general ideas from physics crop up in our field and bear good fruit. Ideas about global operation of nerve nets are inspired by the theory of oscillations. The idea of cell assemblies is related to gravitational attraction (Aertsen and Gerstein, 1985; Gerstein and Aertsen, 1985) and possibly to explosions. Synfire chains owe much to Huygens' principle of the derivation of ray optics from wave optics. There too, rays can cross unperturbed, and the continuation of the ray from any point along its path depends on the microscopic phase relations.

35.4 Role of the Inhibitory Neurons

Our entire essay up to this point was focussed on a re-evaluation of the system of pyramidal cells, which on the sheer force of their numerical prevalence may be taken as the skeleton cortex. Even the three theoretical models of cortical function which we have sketched in this chapter could do entirely without any consideration of the inhibitory interneurons. Yet, they are there and certainly play an important role even if their number is small. The inhibition which the non-pyramidal neurons most likely exert can be assumed to be relatively powerful because it is less diffuse than the excitation which spreads through the network of pyramidal cell collaterals, and more concentrated on strategic regions of the target neurons, such as the cell body and the initial segment of the axon.

In terms of the general picture of cortical function which we propose, there are three distinct reasons which would lead one to postulate inhibitory neurons even if their presence had not been demonstrated physiologically. The first reason is stability. A network with so many and such diffuse excitatory interactions is perennially in danger of exploding into a condition of global, senseless activity. The frequency of epilepsy in the human population shows how real this danger is, and how close to pathology the normal operation of the brain. Although there probably are extra-cortical mechanisms which keep the amount of cortical activity within reasonable bounds, the presence of inhibitory neurons in the cortex may well be necessary to prevent local maxima of activity from spreading.

The second reason is apparent especially in the areas of the cortex concerned with sensory input. The efficiency of "lateral inhibition" in enhancing signals and suppressing irrelevant information was shown in visual, auditory and somatosensory perception. Some of the perceptual effects ("Mach bands" etc.) may be due to inhibitory interactions in the cortical areas which receive the sensory input in the form of orderly geometrical projections.

There is a third reason, connected with the hypothesis of Hebbian learning, for which inhibitory interneurons are a necessary implement of the associative network. A "Hebbian rule" which makes synapses grow stronger when the neurons which they connect fire in synchrony is not beyond imagination. Such a rule leads naturally to the learning of events or situations characterized by the joint occurrence of a number of elementary constituents, as we have seen. However, we are also able to learn events or situations for which the *non*-occurrence of a certain elementary event is a necessary condition. The printed letter P is defined by the absence of any addition which would turn it into an R or a B, as well as by the features which it has in common with these characters. In a similar way most concepts are probably represented in the brain by a combination of some positive and some negative properties. To amend the Hebbian learning rule in such a way as to make the association of negated and non-negated terms possible would imply synaptic mechanisms far more complicated than the ones we are prepared to assume at present. It is much more likely, then, to assume that only positive associations are learned in the Hebbian way, and that all the terms involved in the association are represented by pyramidal cells. But some of these receive their input via inhibitory interneurons and can participate in the cell assembly only when their input is not active. Thus the establishment (and activation) of a cell assembly may depend on the conjunction of both positive and negative terms, and there is no need to complicate the elementary mechanism of plasticity in the synaptic membranes. All we need is a population of inhibitory interneurons interspersed between the cortical input and some of the cortical neurons, and between the cortical neurons themselves.

The anatomy fits this idea. The laminar distribution of both cell bodies and terminations of GABA-immunoreactive (=inhibitory) cells in the primate visual cortex follows closely the laminar distribution of geniculo-cortical terminals (Jones et al., 1994).

We shall discuss in the next chapter an application of this idea to well-known facts of visual physiology.

The interpretation of the cerebral cortex as "nothing but" a memory store that operates in a statistical fashion and is built accordingly, may not be convincing to a majority of brain scientists. Objections come from those colleagues whose imagination is based less on neuroanatomy than ours, and more on physiological (and psychological) facts that would agree better with a scheme akin to *algorithmic computation* so familiar from digital computers. We shall meet them on their own grounds and directly invade two of their strongholds, the organization of the visual cortex (Chapter 36), and neuronal mechanisms of language (Chapter 37). We are convinced that the controversy between algorithmic and statistical theories of brain function will eventualy be resolved in a synthesis of complementary facts and ideas.

Chapter 36

Feature Detectors and Orientation Columns

In the first of the two concluding chapters we shall try to make our essay less extravagant by relating it to more familiar concepts of cortical physiology. We intend to show that the statistical nature of cortical "wiring" proposed by us is well compatible with known facts from visual physiology (Hubel and Wiesel, 1977), even if the very same facts have sometimes been taken as proof against it. The gist of our argument is that slight variations of the statistics of neuronal connections such as might be produced by variations in the shape (and relative position) of cortical neurons may have unsuspected electrophysiological consequences. To make this plausible, we first want to acquaint ourselves with the range of variation of neuronal shape in Golgi pictures.

This is exemplified for pyramidal cells in Fig. 84, extracted from Cajal 1911. The pyramidal cells in the upper left are what we might call the standard shape. Although the sample is from the human brain (first temporal convolution), their appearance is much like that of the pyramidal cells in the mouse cortex, illustrated for example in our Fig. 79. To the right are some pyramidal cells from the human pyriform cortex with very densely ramified basal dendrites, in striking contrast to their "normal" apical dendrites. The total length of the basal dendrites may not be larger than in the "standard" type, but the relative dendritic density in the sense of our Chapter 19 is undoubtedly much higher. The pyramidal cells C, D, E at the lower left, taken from the human insula cortex, are remarkable because of their basal dendrite initial segment which breaks up in many branches only at a certain distance below the cell body. There are other pyramidal cells (A, B) in the same region which have a rather normal appearance, so that the effect cannot be entirely due to any local, possibly mechanical influence. Finally, the large pyramidal cells at the bottom right belong to the human visual cortex and are distinguished by their long, horizontally running basal dendrites. These are the well-known "Meynert cells".

Figure 85 shows similar variations for non-pyramidal cells. The cells marked b and d in the upper left picture, those marked D and E (upper right) and d (bottom right) are stellate cells with dense axonal trees, but the shape of their axonal ramification varies greatly, some being rather spherical, others elongated vertically (E upper right, d bottom right) or composed of straight vertical branches (d, upper left). Some of the neurons in this sample correspond to our class of Martinotti cells: B and C (upper right), A, B, C, D (lower left).

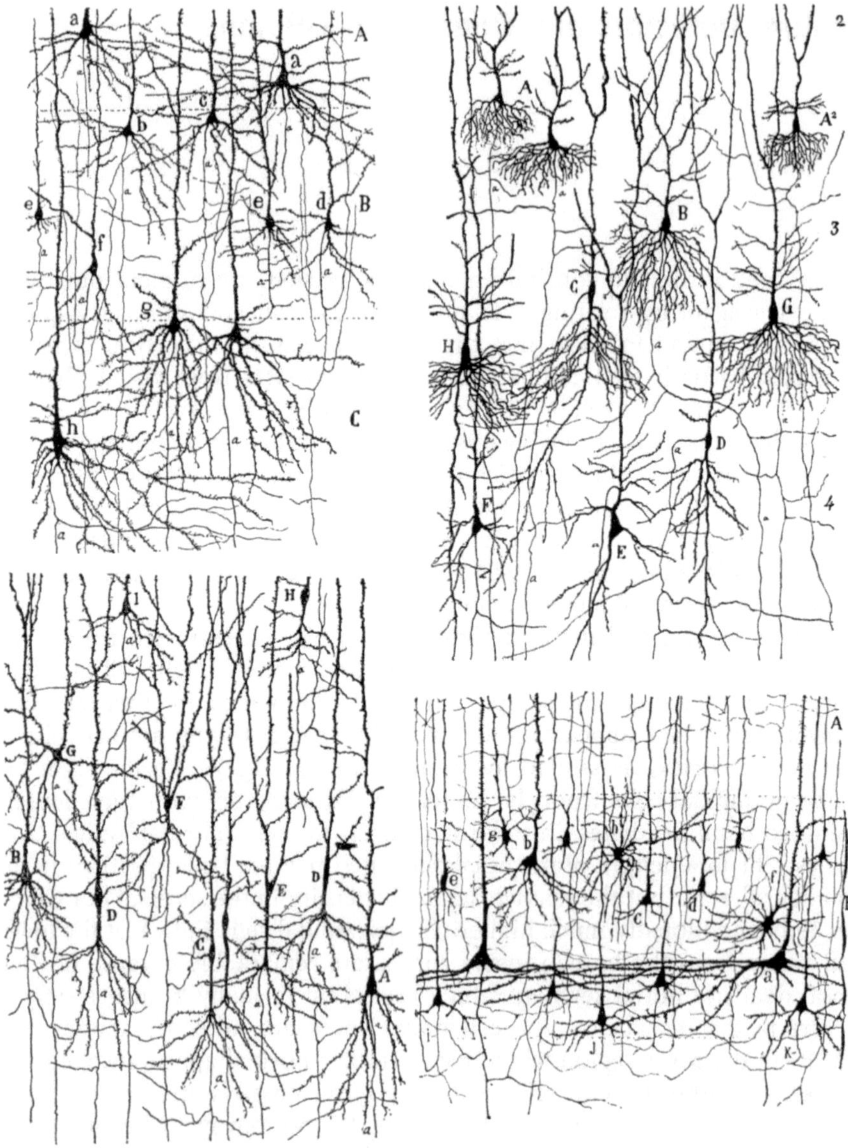

Fig. 84. Pyramidal cells of different shape from the human brain: temporal cortex *(upper left)*, pyriform cortex *(upper right)*, insula *(lower left)*, visual cortex *(lower right)*. Composed out of four illustrations in Cajal (1911)

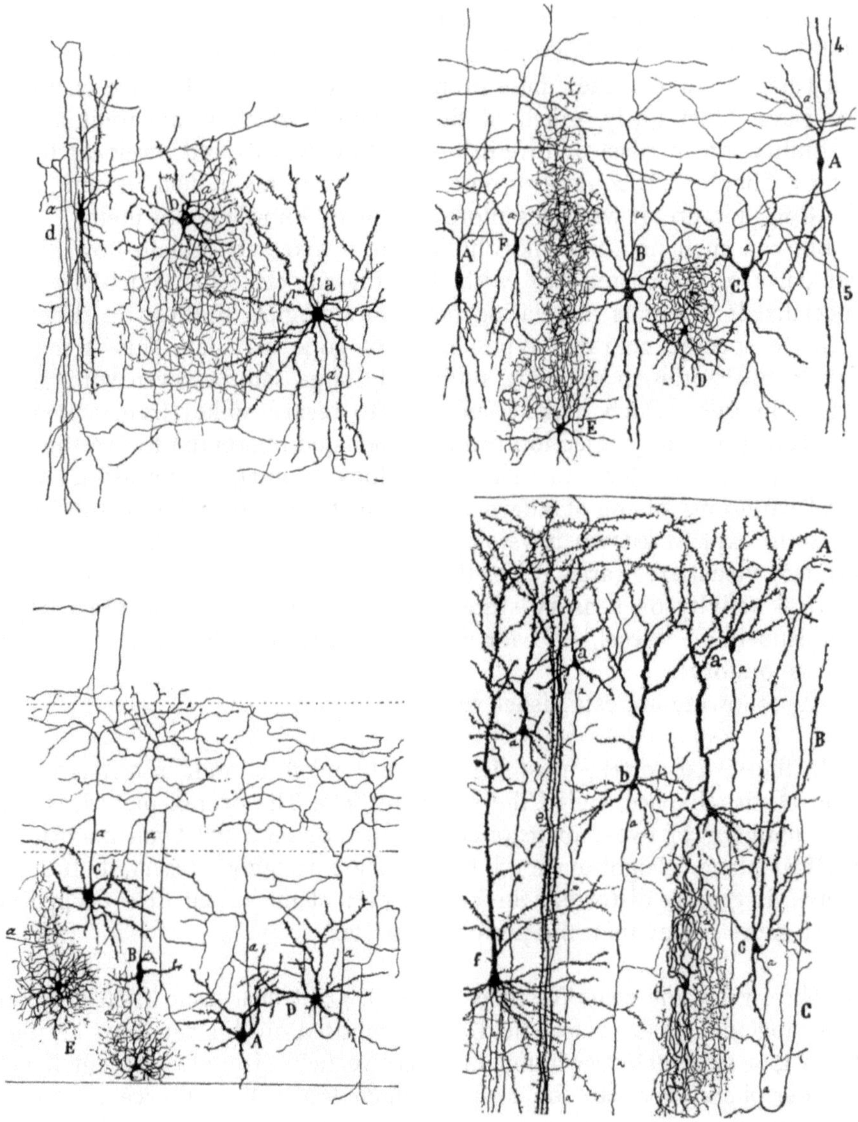

Fig. 85. Various kinds of stellate cells. The size and shape of their axonal ramification varies (spherical *b*, *upper left*, *D*, *upper right*; elongated *E*, *upper right*, *d*, *lower right*), also the "style" of the ramification (mainly vertical branches *d*, *upper left*). Some of these neurons are Martinotti cells in our terminology (e.g. *B*, *upper right*, *A*, *B*, *C*, *D*, *lower left*). Composed out of four illustrations in Cajal (1911)

The different geometry of these neurons certainly influences the probabilities with which neighbouring neurons are connected (in the sense of our Chap. 20) and makes these probabilities depend on distance in different ways. Most likely, much of the variation which is apparent at low magnification in Nissl or myelin preparations (cyto- and myelo-architectonics, Chapter 29) could also be explained in terms of such variations of the Golgi picture, if we had a more systematic knowledge of them.

But can such neuronal morphology be invoked to explain how different parts of the cortex deal with the different tasks for which they are specialized? Attempts at relating neurophysiological findings in different areas to specific neuronal types have not gone very far. This is mainly due to difficulties in securing sufficiently large samples even with modern techniques which make it possible to record spikes from a neuron and later stain it by injection through the same micropipette. We think it is legitimate, therefore, to relate physiology to anatomy by asking what local variations of the standard cortical histology must be postulated in order to explain known physiological findings. We shall do this for the primary visual area (area 17) of the monkey for which the dependence of cortical activity on sensory stimuli has been worked out in great detail (see Hubel and Wiesel, 1977 for a review of much of the early work from that laboratory).

The following are established facts:

a) In the visual cortex, most neurons respond to stimuli in a narrow region of visual space, called the receptive field. The overall picture is that of a *mapping* of two-dimensional visual space onto the two-dimensional cortical plane. Neighbourhood relations are roughly preserved, but the metric is grossly distorted, the central part of the visual field occupying much more cortex than the peripheral parts.

b) The *size of the receptive* fields varies a great deal even when they are close together in the visual field. Superimposed on this random variation there is a systematic variation of receptive field size with "eccentricity": the peripheral ones are on average much larger than the central ones. This effect can be entirely related to the metric distortion of the overall projection: if the receptive field size (in degrees of visual space) is projected back onto the cortex, their size (in mm cortex) is fairly constant irrespective of their location.

c) The *region of cortex corresponding to the receptive field* of one neuron measures about 1 mm across. Such a region contains about 200 000 neurons and receives about 400 (according to a different estimate 1 000) afferent fibres, each corresponding to one point of visual space. The mechanism which conveys information from the receptive field to a single neuron in its interior must be contained in such a 1 mm region of cortex.

d) Most neurons in the visual cortex have *anisotropic* receptive fields. The response of the neuron is absent or very weak for diffuse illumination of the receptive field and also, in most cases, for point-like stimuli. It is maximal for an elongated patch (white on black or black on white) whose long axis has a certain orientation in the visual field, different for different neurons: cortical neurons are "astigmatic". The astigmatism must arise in the cortical mechanism that is responsible for the receptive fields. The response of the neuron in many cases depends on the visual stimulus moving through the receptive field in one direction or the other, or in either direction, always at right angles to the long axis of the stimulus configuration.

e) When the neuron responds to stationary stimulus configurations, it is often possible to define subfields of the receptive field for which the response of the neuron varies in opposite directions: the neuron may be inhibited by a white line in one part of its receptive field, and excited by the same white line positioned in another part of it. Neurons with this property are called *"simple cells"*, and the others *"complex cells"*.

f) The "orientation", i.e. the axis of the astigmatism, depends strongly on the location of the neuron in the cortex. Within a narrow region of, say, 50 μm cortex, all neurons are likely to have receptive fields with orientations differing by no more than 30°. If the dependence of the orientation on the location of the corresponding neurons is tested along straight lines across the cortex over stretches of several hundred micrometers, a high degree of regularity is revealed. As the electrode proceeds across the cortex testing one neuron after another, the orientations in the visual field turn clockwise or counterclockwise, at a rate of about 360°/mm. There may be occasional breaks in the sequence, or the rotation may change from clockwise to counterclockwise. Orientation never seems to remain constant over long stretches of cortex. There is a tendency, however, evident in many records, for orientations to repeat for neurons about 0.5 mm apart, even if the rotation has changed from clockwise to counterclockwise in between.

It was this puzzling layout of neurons with differently oriented astigmatism which prompted us to postulate a mechanism in terms of neuronal interaction. The first step (Braitenberg and Braitenberg, 1979) was an analysis of the published records with the expectation that they would perhaps give away the two-dimensional map of orientation-selective neurons on the cortical surface. The main idea was that an arrangement of orientations around centers would be the most natural way of introducing anisotropy into an otherwise isotropic network. It was indeed possible (Fig. 86) to interpret the records in this way. If there are centers in the cortex around which the neurons are arranged, it seemed that the map of their preferred orientations, projected onto the cortical plane,

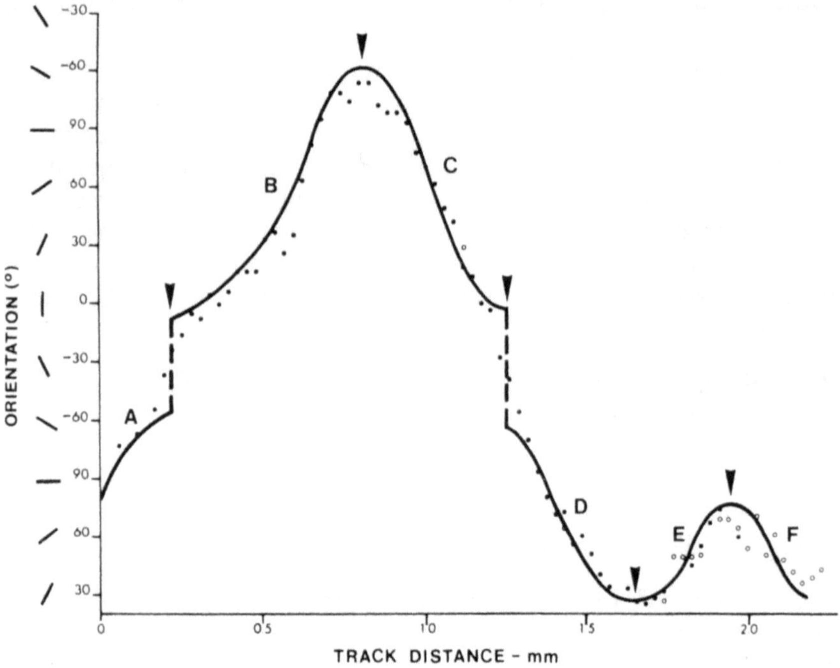

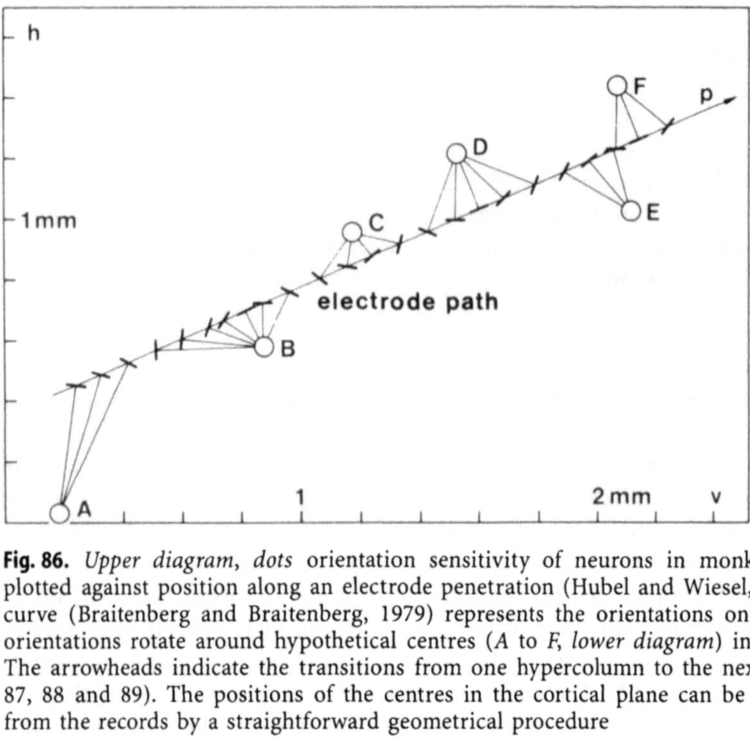

Fig. 86. *Upper diagram, dots* orientation sensitivity of neurons in monkey area 17 plotted against position along an electrode penetration (Hubel and Wiesel, 1977). The curve (Braitenberg and Braitenberg, 1979) represents the orientations one expects if orientations rotate around hypothetical centres (*A* to *F*, *lower diagram*) in the cortex. The arrowheads indicate the transitions from one hypercolumn to the next (see Figs. 87, 88 and 89). The positions of the centres in the cortical plane can be determined from the records by a straightforward geometrical procedure

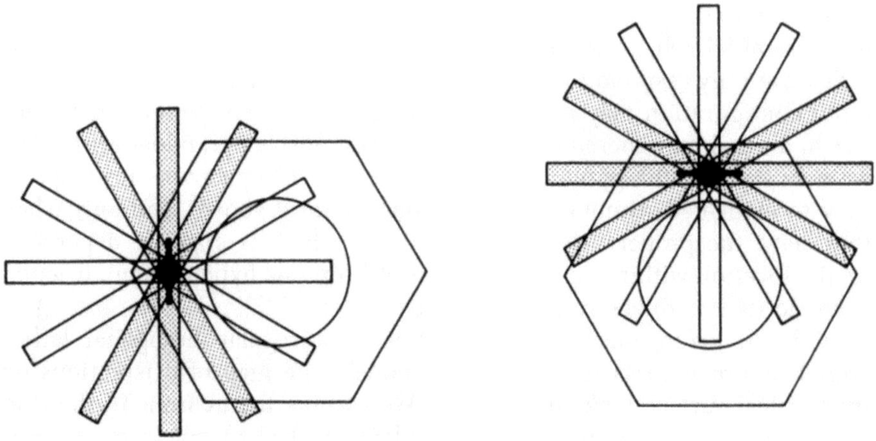

Fig. 87. To explain how a cortical neuron *(black dot)* becomes a detector for lines *(elongated rectangles)* of a certain orientation *(stippled)* thanks to its being under the influence of inhibitory neurons located in the centre *(circle)* of the hypercolumn *(hexagon)*. Both a more extended stimulus and a line of the wrong orientation *(white rectangles)* activate the inhibitory centre and hence prevent the neuron from responding. Depending on the position of the neuron within the hypercolumn (compare the two diagrams on the *right* and on the *left*) the neuron has different orientation preferences (vertical on the *left*, horizontal on the *right*)

could well be such that all orientations lie on circles around the centers. Moreover, it turned out that the centers that had to be hypothesized in order to fit the records were distributed over the cortex at fairly regular distances of about 0.5 mm one from the other.

At the time nothing was known in the histology of monkey area 17 that repeated at the predicted distances. Shortly afterwards (Humphrey and Hendrickson, 1980; Horton and Hubel, 1981) "cytochrome oxidase blobs" were discovered, regions of high metabolic activity that formed a regular pattern of dots over the striate cortex, with a spacing that agreed remarkably well with our prediction. These could well be the singularities which we needed and we were therefore encouraged to produce a more concrete model (Braitenberg, 1983, 1985, 1986).

Imagine a tesselation of the cortex in hexagonal regions called *hypercolumns*. In the centre of each hypercolumn (Fig. 87) there is a circular region in which we suppose inhibitory neurons to be concentrated. Their inhibitory influence is strong and is limited to the hypercolumn in which they are situated. In the remainder of the hypercolumn, pyramidal cells of the ordinary kind predominate. Consider one of the pyramidal cells on one side of the hypercolumn (Fig. 87, left). Let an elongated stimulus configuration (a "bar") be projected through the input fibres on that pyramidal cell (black dot).

The neuron can only respond if the long axis of the bar does not deviate too much from the vertical (stippled bars), since otherwise it would excite the central inhibitor (white bars) which would in turn block the response of the pyramidal cell. A pyramidal cell above the inhibitory centre (Fig. 87, right) can only respond to more or less horizontal bars (stippled). A larger input configuration may not fit into the space between the inhibitory centres at all and may therefore not find any responsive pyramidal cells. Thus we may say that the pyramidal cells of the area 17 hypercolumns respond only to elongated stimulus configurations (point d above), and only when they have the proper orientation (point f), which is strongly dependent on the location within the hypercolumn. Within one hypercolumn, orientations do indeed rotate around the centre.

With an arrangement of the centres in a regular hexagonal lattice (which is not necessarily part of the model), the map of orientations on the cortical surface looks like Fig. 88. We assume the neurons in the centres of the hypercolumns (e.g. in the place marked *a*) not to be orientation-sensitive. Going once around the centre, say around *a*, every orienta-

Fig. 88. Assuming a hexagonal array of hypercolumns of the kind illustrated in Fig. 87, the distribution of orientation-sensitive neurons on the cortical plane would form a pattern as shown here. The *Saturn-like symbols* represent neurons, their preferred orientation in the visual field is indicated by *the line crossing them*. The layout of the neurons is in cortical coordinates, the corresponding preferred orientation in visual coordinates, but they can be represented on the same diagram because of the well-known projection of visual space on the cortical plane. *a* is the centre of a hypercolumn; each orientation occurs twice in the neurons surrounding it. *b* is a corner between hypercolumns which is also surrounded by neurons of different orientation but each orientation occurs only once

tion occurs twice. But there are other places, the corners between hypercolumns (*b*) around which orientations also rotate, each orientation occurring only once in a full circle.

Figure 88 is misleading in that it may suggest that the oriented receptive fields belonging to single neurons are the size of the small Saturn-like symbols which mark their orientation. As a matter of fact, as we have said before (point c above), receptive fields are much larger, about two or three times hypercolumn size, if they are measured in cortical coordinates. This makes them larger than the mechanism that determines their orientation, which is puzzling. Our answer is that what the physiologist measures as the receptive field of a single neuron, is in reality a composite receptive field of several neurons located in different hypercolumns but all having the same orientation. If we bear in mind that most neurons in the cortex are pyramidal cells and that these tend to unite into cell assemblies when they are often active together, we should expect neurons of the same orientation to exert excitation onto each other within a certain region of the cortex. With all the oriented visual stimuli normally sweeping the visual field, they have often been activated in synchrony or in close succession and their reciprocal excitation has thereby been strengthened in the Hebbian way. When the physiologist records the responses of a single neuron to visual stimuli, he may be observing in reality the responses of a cell assembly of which the particular neuron is a member.

Thus we may consider stimulus configurations, the size of receptive fields, and their effectiveness in exciting cortical neurons (Fig. 89). A is a large stimulus which remains ineffective. B and C are bars which can produce responses in only some of the neurons they touch. These may still have united into a cell assembly, and B and C may therefore be taken as stimuli which excite the same receptive field. But there is no position in which one of the bars could have only excitatory, or only inhibitory effects. The physiologist would then call the region in which B and C are effective a "complex field". On the contrary, D, E, F are three positions of a bar which may correspond to two inhibitory and one excitatory subfields of a "simple cell" (point e above). Note that according to this scheme, simple and complex cells should have different orientations: simple cells the orientations corresponding to rows of hypercolumns, complex cells the other orientations. This is a strong prediction of our model. The bar marked G on Fig. 89 represents the situation which Hubel and Wiesel called "hypercomplex cell". What is characteristic for these is that the response disappears when the stimulus bar becomes too long. It is plausible in the case of bar G that if its length were increased it would reach into the inhibitory centres at either end, and the response of the corresponding assembly would be at least diminished. It is also clear that the hypercomplex effect would be most obvious in cells with very short fields, which might be completely silenced by inhibition from both ends.

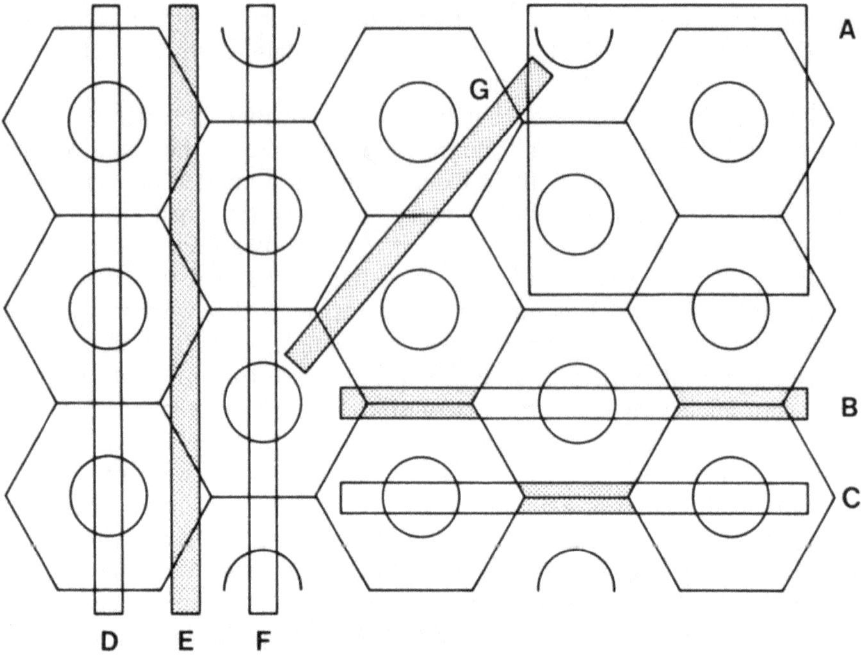

Fig. 89. The receptive fields of Hubel and Wiesel (1977) are much larger than the hypercolumns, if they are projected onto the cortical plane. They are defined by position, length and orientation of lines which activate single neurons. If the cortex is composed of hypercolumns as in Fig. 87, it can be seen that different lines may excite entire rows of cortical neurons *(E)*, or inhibit entire rows of hypercolumns because they go through their inhibitory centres *(D, F)*, or excite neurons in some hypercolumns but not in others *(B, C)*. We suggest that the orientation of lines parallel to rows of hypercolumns, such as *D, E, F,* is what defines so-called simple cells. A line such as *G* may activate neurons in four hypercolumns provided it is not too long, otherwise only in two hypercolumns. A stimulus *(A)* which does not fit into the space between inhibitory centres does not activate any cortical neurons except the inhibitors. It is assumed that neurons of the same or of neighbouring hypercolumns, which are often excited by the same line in the visual field, tend to unite into a cell assembly. The assembly of neurons exciting each other defines the size of the receptive field for every neuron of the assembly

There have been various attempts at experimentally verifying the geometry of "orientation columns" by either staining active neurons with the deoxyglucose technique after appropriate stimulation (Löwel et al., 1987), or by observing their activity in the living brain by "optical recording" from the surface of the cortex (Blasdel und Salama, 1986; Grinvald et al., 1986; Grinvald and Malach, 1994; Das and Gilbert, 1997). In all cases, a patchy pattern is seen when only one orientation is presented in the visual field, and the period of the pattern corresponds rather well to the spacing of the hypercolumns as postulated by us. The experiment by Löwel et al. (1987) was especially designed to test the hypothesis of a centric arrangement. Fig. 90 shows what we expect to happen in the array of neurons of

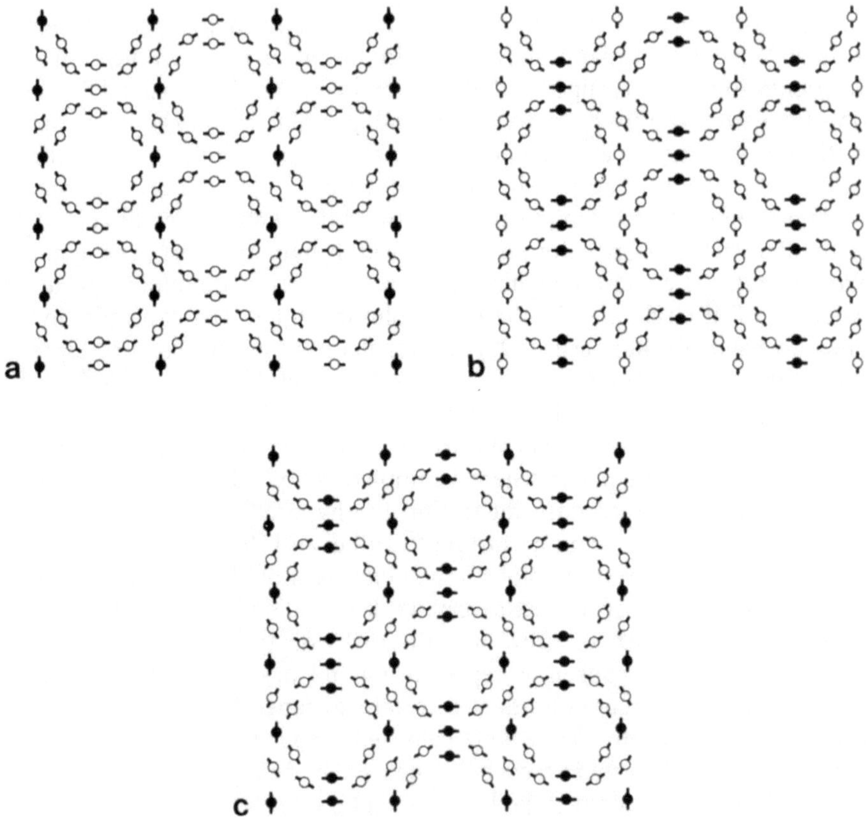

Fig. 90 a–c. The expected outcome of an experiment by Löwel et al. (1987). If in the array of Fig. 88 only the neurons responsive to vertical lines are activated, one expects the outcome of an activity stain *(black symbols)* as in the *upper left* diagram *(a)*. For only horizontal lines one expects something like the *upper right* diagram *(b)*. If vertical and horizontal stimuli are combined, the pattern should be as in the *bottom* diagram *(c)*. This is very similar to the experimental results

Fig. 88 if (a) only vertical stimuli, or (b) only horizontal stimuli or (c) vertical and horizontal stimuli are presented in combination. The diagram shows the neurons that would be stained by an activity stain (like deoxyglucose) in the three situations a, b and c. In c there would be twice as many rows of stained elements as in a and in b, but the number of patches along each row would be unchanged. This is exactly what Löwel et al. (1987) find in their brilliant deoxyglucose experiment, although curiously they take their evidence as proof for the opposite hypothesis, i.e. for orientation columns in the form of stripes, as originally postulated by Hubel and Wiesel, rather than in any centric arrangement.

All the experiments in which the preferred orientations of cells are mapped onto the surface of the cortex by the technique of optical recording (the loci of maximum activity in the cortex are determined separately for variously oriented stimuli in the visual field and then superimposed to form an "orientation map") show places in the cortex in which the entire set of orientations are arranged radially around centers. These so-called "pinwheels" were taken by some (e.g. Grinvald and Malach, 1994) to be confirmation of the centric arrangement postulated in our model, and by others (Götz, 1987) as proof against it, since in the pinwheels each orientation appears only once in one complete circle around the center, whereas according to us diametrically opposite positions in each hypercolumn (a in Fig. 88) should prefer the same orientation, each orientation therefore appearing twice in a full circle around the hypercolumn center. The contradiction is only apparent (Braitenberg, 1992) and results from an incorrect interpretation of the orientation maps in terms of our model. If on the maps (Blasdel and Salama, 1986; Grinvald et al., 1986; Grinvald and Malach, 1994; Das and Gilbert, 1997) on which preferred orientations are represented by different colours one seeks the places where a complete set of colours comes together in one point, one is certain not to find the hypercolumn centers. In fact, there, according to us (and according to Hubel and Wiesel, if the hypercolumns are centered around "blobs") the neurons should not have any orientation tuning at all, or at best a very weak one. The hypercolumn centers should show as black, or grey, or white disks on the coloured maps (depending on what is chosen as representing "no orientation"), surrounded by a crown of different colours at a certain distance from the centers. Indeed, for all we know, the various coloured maps which do not show the "no orientation" loci must have been manipulated in some way. The places where all the colours come together in one point, i.e. where neurons responsive to different orientations are housed in a narrow region, almost certainly correspond to the *corners between the hypercolums* (b on Fig. 88), and there indeed each colour (every orientation) appears only once, as can be easily seen on Fig. 88. In the corners, according to our model, and according to experiments (Das and Gilbert, 1997), the tuning of the neurons should be sharp. It comes as no surprise that the centers of the pinwheels do not correspond to the "blobs" (Grinvald and Malach, 1994), which we locate in the centers of the hypercolumns, for the pinwheels are not centered on the hypercolumns. What is surprising is that the discontinuities in the orientation map, which one would expect at the borders between hypercolumns, also tend to coincide with discontinuities, or non-linearities, in the visuotopic map (Das and Gilbert, 1997), as if the pixels (= picture elements) which belong to one hypercolumn were somehow interpreted by the cortex as belonging to the same point of visual space.

We have taken some time in describing experiments and ideas that might seem foreign to the main theme of our essay. In reality, they are relevant. If the model of orientation columns which we propose, or any model of a similar kind proves to be correct, we may take this as yet another argument in favour of our statistical view of cortical connectivity. In our model of area 17 there was only *one small departure* from the overall picture of cortical connectivity which we proposed in the main part of our essay, namely the *assumption of lumped inhibitory neurons* in patches of the visual cortex. There was no assumption about any specific wiring within or between hypercolumns, beyond the general assumptions about the plastic excitatory synapses in the "skeleton cortex" and the inhibitory interneurons. And yet most of the facts we have listed (a to f above) can be comfortably explained by the model we proposed (Braitenberg, 1985, 1986), including the size of receptive fields, the difference between simple and complex cells, the scatter of the localization of receptive fields (which has something to do with the statistical nature of the pyramidal cell to pyramidal cell wiring) and the relative fixity of their orientation.

Unfortunately, one of the assumptions on which our model of the orientation mechanism rests, the concentration of inhibitory neurons in the hypercolumn centers, has not yet been confirmed in the experimental literature. If the centers are the cytochrome oxidase blobs, we should find more inhibitory neurons there than in the surrounding regions, but this is not the case (Fitzpatrick et al., 1987; Hendry et al., 1987). However, there are several different kinds of GABA-ergic inhibitory neurons, distinguished by an entire battery of histochemical and immunoreactive markers, and very likely subserving different functions in the cortex. There is hope: one group of GABA neurons, characterized by immunostaining with the monoclonal antibody Cat-301, are indeed preferentially localized within the cytochrome oxidase rich regions of the monkey cortex (Hendry et al., 1984). Our model requires a set of special inhibitory neurons with dendrites housed in the hypercolumn centers and axons emanating radially from there to reach the surrounding pyramidal cells. Whether such an arrangement exists, is the crucial question awaiting an experimental answer.

Our basic tenet, however, was that the orientation sensitive neurons are arranged in an orderly manner around certain periodically spaced singularities in the visual cortex. This now seems to be generally accepted and experimental confirmation is abundant. After simulation of the visual field with lines of one orientation, an "activity stain" always produces a spotty pattern on the surface of the cortex, and the localization of the spots is different for different orientations. Today, no one would call a row of spots an "iso-orientation band" or "slab" (Löwel et al., 1987; Gilbert and Wiesel, 1989; Löwel, 1992) any longer. Iso-orientation *spots* appear in experimental studies on horizontal fiber connections in the visual cortex of cats and monkeys, where they seem to induce clustered fiber terminations, as well as connections between different spots associated with one and the same orientation (Gilbert and Wiesel, 1989; Amir et al., 1993; Malach et al., 1993).

Not everybody is convinced by our model. For a critical review see Erwin et al., 1995. For an attempt to relate functional models to realistic cortical anatomy see Lund et al., 1995.

Chapter 37

Human Language: The Ultimate Challenge for a Theory of the Cortex. An Essay

37.1 The Peculiar Phenomenon of Language

By emitting air from the lungs, opening, closing or constricting the wind channel in various places, setting the vocal cords in vibratory motion at various frequencies or letting them loose, we are all able to produce sequences of sounds which induce particular mental states in the listener and inform him about our own. We also learn to code the audible utterances into sequences of graphic symbols when we write, and to decode the writing again in terms of spoken language. All this requires an unusual degree of motor and sensory finesse and, in particular, an ability to generate and to recognize temporal order in long sequences of symbols, an ability which is perhaps peculiar to humans.

If the cerebral cortex is involved in this, we must show how the functions we ascribed to it apply to language, and what additional mechanisms would have to be postulated to explain the *grammar of the brain.* We shall summarize here some ideas developed in our group over the past years (Braitenberg, 1980; Braitenberg and Schüz, 1992; Braitenberg and Pulvermüller, 1992; Braitenberg, 1996).

37.2 A Hierarchy of Discrete Elements

Language is not a continuum of sounds, nor is it a continuum of grammatical categories or of meanings. At various levels of complexity we discover discrete sets of elements, the activation of each member of a set excluding the simultaneous activation of another member of the same set. The most impressive set of linguistic elements, the *alphabet,* has actually become the paradigm for, and has given the name to, all sets of mutually exclusive symbols.

1) Every language in the world can be written with sufficient degree of accuracy by means of a keyboard of about 30 keys. The keys correspond to letters of the alphabet, and the letters roughly correspond to the elementary sound patterns (or better, episodes) of speech which the linguists call *phonemes.* To be sure, at the level of the acoustic signal it is not always easy to determine the boundaries between successive pho-

nemes, which are often slurred and may even influence each other's sound pattern. The successive activation patterns of the musculature responsible for the generation of phonemes may be equally slurred. Still, there is plenty of evidence to show that the *intended* phoneme is always one particular member of the set, and the phoneme *perceived* by the listener is also unambiguously a particular one (not necessarily the same one as intended by the speaker when the transmission is poor). This implies that the "keyboard" responsible for the discreteness of the alphabet of phonemes resides in the brain rather than in the motor or sensory periphery. The actual phonemes placed on the keys of the keyboard depend on the language first acquired by the individual. Some phonemes are common to all languages, but others vary, as one discovers when trying to learn the correct pronounciation of a new language in later life.

2) Spoken language is a succession of episodes, called *syllables*, each marked by a maximum of acoustic energy in the speech signal. Syllables are produced at a rate of four or five a second in all languages. Although each syllable is uttered (and understood) as an indivisible unit, and can in fact hardly be interrupted once started, syllables can be described as successions of phonemes, their number ranging from one to about five per syllable. Of all the possible successions of phonemes (of the order of tens of millions for up to five-letter syllables), only a small fraction (of the order of a few thousand) in any particular language actually occur as syllables.

3) *Words* are the units of language most familiar to us, although they are the most elusive ones in terms of an exact definition. Words correspond to concrete objects, such as *milk* or *house*, but also to much less tangible things, like *completion* or *meaning*. The typographic form of the word, as a segment of text between two spaces, does not help to identify words, since spoken language utterly disregards the boundaries between words. Besides, if *medusa* is one word, is *Portuguese man-of-war* also one word? Are *dog* and *dogs* different words, or just variations of one and the same? Are *run* (to move in a certain way) and *run* (a *run* in the stocking) the same word or two different words?

4) Akin to words, but more useful in the analysis of language, are *morphemes*. These are the components of language to which a meaning is associated (phonemes carry no meaning, although they discriminate between different meanings, as the *n* discriminates between *bar* and *barn*). Morphemes are also the elementary, not further divisible components which result from a grammatical analysis of language. A morpheme may correspond to one word (*tree, walk, good, but, the*) or to part of one word (the plural morpheme *-s* in *trees*, the past morpheme *-ed* in *walked*) or even to more than one word (*in that* meaning because, *take after* meaning resemble). Most morphemes correspond to single syllables, but there are many which are represented by polysyllabic words. Some

are incorporated in syllables by changing their form: the past-morpheme making *ran* out of *run,* the plural morpheme changing *man* into *men.*

The meanings represented in language by morphemes are manifold. Obviously, nouns are different in kind from verbs and adjectives, abstract nouns are different from concrete nouns, proper nouns from common nouns, adverbs are something else again, as are prepositions and conjunctions. There is one distinction which overrides all the others. Some morphemes, sometimes called *lexical morphemes* refer either concretely (*green, house, jump, slow* etc) or more abstractly (*object, achievement, intend, good* etc) to things and events in the world. Others, which we may call *grammatical morphemes* (e.g. the past-morpheme *-ed,* the articles *the, a, an,* prepositions like *to, in,* conjunctions like *and, but*) carry no meaning in the conventional sense, but provide the connective tissue which makes a phrase out of a string of words. What they represent refers only indirectly to the world, and it does so only by modifying the meaning of the lexical morphemes to which they are associated. (The terms *content words* and *function words* are often used to make the same distinction, as well as *open class words* and *closed class words*).

5) Certain strings of words are legitimate in language and others are forbidden by the rules of *grammar.* When the speaker applies these rules correctly, he sometimes may do so "by ear". Having heard often enough that *a rose smells* but *roses smell, a bell rings* but *bells ring,* he will probably always put the *s* in the right places and never say *a rose smell* or *bells rings,* or by extension, *horses runs* or *the rule apply.* In some (e.g. slavic) languages "grammar by ear" may be more prominent, but in English it is surprisingly difficult to detect any such superficial rules. And that they are not truly superfical, e.g. in the sense that in the examples above the *s* is either with the noun or with the verb, becomes apparent when one considers *men run,* or *oxen run,* or *a man must.* Evidently, what correlates with the plural form of the verb *run,* is not so much the plural marker *s* of the noun *horses,* but some abstract representation of the plural which is activated also with the word *oxen* or *men.* Thus, besides an alphabet of phonemes and an inventory of morphemes of various kinds, we must postulate a set of *grammatical categories* which in their interplay with the overt words of the language enter the rules of grammar. The production (and perception) of language is a process that involves parallel activity at three levels, that of the imagery which corresponds to the meanings expressed, that of the strings of words and that of the grammatical categories to which each of the words belongs.

Like *phonemes* and *morphemes,* the *categories of grammar* form sets of discrete, mutually exclusive elements. Between *singular* and *plural,* between *male* and *female* pronoun, between the *first, second* and *third person,* the *present, past* and *future tense* of the verb there are no transitions or mixed forms. In recent times, the science of grammar became more and more complex the more it seemed that computer science would unravel its complexities, and new categorizations were introduced continually. However, the traditional categories of school grammar are still valid and may suffice for our purposes.

37.3 The Neural Embodiment of Linguistic Elements

Phonemes. The production of a particular phoneme ("pronouncing a letter") requires a complex pattern of muscular activations in the larynx, pharynx, tongue and lips. There is one such pattern for each phoneme of the languages the person has learned to speak. The time it takes to utter a phoneme is from about 20 to several hundred ms. Within that time, some phonemes have temporal structure, in the sense that the time between the activation of various components of the motor pattern (e.g. opening the lips and setting the vocal cords in vibratory motion) is critical and may distinguish one phoneme from another. When the production of the phoneme is completed, the activation of the various muscular components must give way to a different pattern of muscular activations required for the production of the next phoneme.

All this is compatible with the idea that the neurons controlling the motor pattern for each phoneme are organized in something like a *synfire chain*, an ensemble of elements connected together in an unidirectional way so that activity started at one end activates all the elements in a fixed temporal pattern and then comes to an end. If we assume synaptic delays of one or a few ms, the synfire chain responsible for a phoneme would have to contain something like 50 elements, or more likely, 50 sets of neurons serially arranged.

But this is not the whole story. The representation of phonemes in the brain must also comprise their sensory aspects, i.e. the various acoustic patterns which are heard when each of them is pronounced. There must be groups of neurons somewhere which respond specifically to each of the phonemes, neatly separated from each other, presumably by inhibitory links, each with some internal wiring which matches the temporal aspects of the phonemes, probably again of the synfire chain kind.

For all we know of the topography of motor and sensory systems in the cortex, it is very likely that the motor representatives of the phonemes are housed somewhere in the precentral region, presumably near the lower end of the precentral gyrus, where the mouth and throat regions of the homunculus are localized. The sensory representatives, on the other hand, should be located not far from the primary acoustic area, from which they receive their input. It is quite likely that the neurons which represent phonemes are contained in the two main areas involved in language, traditionally called the Broca and Wernicke centers, one in the posterior frontal and one in the superior temporal lobe of the left hemisphere.

There must be connections between the sensory and the motor representatives of phonemes. In fact, since we are all able to repeat words or single phonemes which we hear from others, we must postulate specific connections, one by one, from each of the sensory neuronal representatives of phonemes to the neurons organizing the corresponding motor pattern. The Broca and Wernicke centers are a few centimeters apart in the cortex, which means that these connections must be by way of the white substance. Indeed, a fiber bundle connecting the two centers is described in textbooks of neuroanatomy.

It would be surprising if the specific connections between the sensory and motor representatives of each phoneme were due to an inborn "wiring". Babies are not able to repeat words until they have gone through the so-called babbling phase, where they playfully utter all the sounds their vocal apparatus is able to produce, while listening to the acoustic effects which their phonetic exercise entails. By association, they learn the correspondence between motor and acoustic patterns. Only then are they able to reproduce the words which they hear in their environment. If there are enough fibers connecting the Broca and Wernicke regions in both directions, the association will work both ways, and we may envisage the entire set of sensory and motor units belonging to one phoneme as a *cell assembly*. However, it is an assembly of a special kind, being composed of two parts which individually have the nature of *synfire chains*. The whole ensemble, once activated, does not stay active but rapidly exhausts its activity, contradicting the nature of cell assemblies which are supposed to hold their activity for some time.

Like some kinds of early learning described by animal ethologists ("imprinting"), and like the establishment of orientation sensitive feature detectors in the visual cortex (Wiesel and Hubel, 1965), the acquisition of the phonemic repertoire actually used in the "mother tongue", which follows the babbling phase, is limited to a "critical period" and is irreversible (Jakobson, 1941). This is the reason why it is so difficult to overcome a "foreign accent".

Syllables. The neural representatives of syllables must be similar to those of phonemes, except that the duration of their activation is greater, since most syllables are sequences of several phonemes. Like phonemes, syllables have a sensory and a motor aspect, probably represented by neurons in the Wernicke and Broca region. Even more clearly than in the case of phonemes, these representatives organize motor output (respond to acoustic input) in a fixed temporal order and must therefore be of the synfire chain type. However, when a syllable is uttered, it is not the initial phoneme, but the whole syllable that is recruited by the speech production mechanism, for many different syllables begin with the same phoneme and recruiting the sequence by its beginning would lead to ambiguities. This shows that the distinction between cell assemblies and synfire chains is not a sharp as we had initially assumed. There must be something that combines characteristics of both, temporal order and rapid extinction as in synfire chains, global activation as in cell assemblies.

The most interesting aspect of syllables is their rhythmic succession. Clearly, the mechanisms governing temporal order within a syllable of 200–250 ms are very different from those which determine the successions of syllables in a word or in a phrase at a rate of four or five per second. Periods of a few tenths of a second, comparable to the duration of syllables, are important in many aspects of behaviour and of perception. "Backward masking" in visual perception is most evident when the interval between two stimuli is about 200 ms. The disturbance of speech production with delayed acoustic feedback (Lee effect) is maximal when the delay is of that magnitude. Most "fast movements" of the arms have a duration of 200 or 300 ms. Most musical notes also. We conclude that much of behaviour, and very likely thought too, is organized in temporal chunks of a few tenths of a second. The mechanism reponsible for this periodic recruitment and updating of information in the cortex may not reside in the cortex itself, but is more likely to depend on some cortico-subcortical interplay. We have already sketched this in a previous Chapter (35) and given it the name of "pump of thoughts".

Lexical morphemes. A morpheme is an entity which ties a meaning to a piece of text, frequently a syllable, sometimes more than one syllable, and sometimes part of a syllable. For morphemes of the lexical kind, the question of their neuronal representation can be answered in a simple way. The *meaning*, roughly corresponding to what is called a *concept* in philosophy, is represented by a constellation of activity in a set of neurons which may be anywhere in the brain. The linguistic aspect of the morpheme, on the other hand, is represented by a constellation of active neurons housed in the language centers of the left hemisphere. The two sets of neurons must be reciprocally connected, so that the word may elicit the concept and, vice versa, the concept may trigger the word.

All of this fits the idea of a *cell assembly* rather well, with one caveat. As far as the *concept* is concerned, we have no difficulty in seeing it represented by a true Hebbian cell assembly, a group of neurons which ignites in its entirety following the activation of any sufficiently large part of it, and, once ignited, remains in the active state for some time. In fact, concepts tend to stick, and they can be evoked by a variety of clues that somehow pertain to them. The difficulty arises with the *word* that is associated with the concept. Its neuronal representation, as we already know, must be a chain of phonemic representatives, rather in the nature of a synfire chain than of a synchronous assembly. Can such a chain ignite an assembly? Does it run into the assembly and trigger an explosion like the fire on the fuse? Are the neurons of the chain part of the assembly? If they are, how do they recover their temporal order when the assembly ignites and triggers the word? Various solutions to these problems come to mind, all possible, perhaps even likely, but still far from being verified experimentally at a neuronal level.

Grammatical morphemes. The other kind of morphemes, grammatical morphemes, the ones which do not refer to facts and events of the outside world, also connect activity pattern of the brain with strings of phonemic representatives, but the meanings of these activity patterns are hard to define. What is the state of the brain that corresponds to the conjunction "and", to the preposition "of", to the definite article "the" or to the past-morpheme "-ed"?

Evidently, if I say "the monkey", rather than "a monkey" or "monkeys" or just "monkey", what is meant by *the* has very little to do with the concept of monkey (in fact I could also say "the donkey" or "the constitution" etc), but rather with a particular function of the word *monkey* in the phrase where it occurs (the *definite article* relates the word to something that has been said already about that particular individual monkey, or to something that will be said later about him). Similarly, little words like *if, is, it, on, of* and many others, most of them one-syllable words and unstressed, have a meaning which can be best defined by reference to the grammatical context which their distribution in the phrase creates. If we look for the neurons whose activity represents that meaning, we would probably find them in the regions of the cortex in which the structure of the phrase is organized, most likely in the speech areas of the left hemisphere.

The different role, and different localization in the cortex of *lexical morphemes* (all over both hemispheres) and *grammatical morphemes* (in the speech areas on the left side) finds good experimental support in EEG-studies on human subjects (Pulvermüller, 1995).

The case is more difficult with some morphemes which, although not being *lexical* in the sense defined above, do not carry a meaning defined in the *grammatical* context either, but one involving more complex functions of the brain. The so-called modal verbs *will, must, may* etc. refer to the emotional setup which underlies the linguistic utterance, and words like *but, although, therefore* make sense only in connection with the ongoing prediction of consequences which operates in the mind of the speaker as well as in that of the listener. We have no way of localizing the neurons which represent the meaning of these morphemes, since we have only tentative ideas about the mechanisms of emotion and of prediction, which almost certainly involve parts of the brain outside of the cerebral cortex.

Grammatical categories. In a sentence such as "The lady I met yesterday at the party expressed opinions on presocratic philosophers which were more revealing about her than about them" there are covert markers for *female* and for *singular* which are triggered by *the lady* and which stay active throughout the phrase, allowing the correct attribution of the pronoun *her* even after 32 intervening syllables, corresponding to a time lapse of about eight seconds. There is another marker for *plural* triggered by *presocratic philosophers* and decisive for the pronoun *them*. There is also a marker for *thing* (as opposed to *person*) that goes with *party* eliminating the attribution of *her* to the *party*. The study of grammar reveals a multitude of such long-term connections within the sentence, sometimes clearly embodied in overt morphemes such as prepositions or word endings, sometimes hidden in the covert mechanisms of grammar.

Where are the neurons localized which, by their activity, set the markers for the grammatical categories, and how are they organized? Their localization is almost certainly in the speech centers of the left hemisphere, for one of the striking effects of lesions in that region is the inability to produce, and to understand, complex sentences even if the ability to produce and to understand individual words is preserved. It is interesting to speculate about the internal organization of the groups of neurons representing grammatical categories. They are almost certainly of the nature of Hebbian cell assemblies. In fact, grammatical categories operate like switches within the sentence which, once thrown in one position, stay put for a (theoretically) unlimited time until they are set back again. For them we must postulate neuronal representatives which have the property of bistability, to borrow a term from engineering: like light switches, which can stay on or off, rather than like the button that operates the door bell which is stable only in the off-position. A cell assembly which, when activated, stays active for some time is a good candidate for such a switch, if we let it have excitatory input to ignite it, as well as inhibitory input to rapidly extinguish it again.

There is an extra bonus in this. Some grammatical patterns in language have a nested structure, for instance when a relative clause is inserted in a sentence and may even contain another inserted phrase. Parentheses (marked by square, round, etc brackets) are a familiar instance of such nesting, and as with parentheses, the various levels of a complex sentence (main sentence, inserted phrase of first, second etc. order) have to be closed in the inverse order in which they were opened. This requires a machine of a special kind, called push down memory, working by the principle of "last in, first out". Cell assemblies can be used to build such a machine, if we assume, quite realistically, that the activity following their ignition slowly decreases in time (the familiar phenomenon of neuronal fatigue, or adaptation, could produce this effect). If there is a cell assembly representing the grammatical marker for *open the first sentence*, and another for *open the first inserted phrase* etc, in the process of completing the sentence, the more recent markers will be more strongly active (less fatigued) than the preceding ones. If the stronger signal is processed before the weaker one, we can easily construct a device which opens and closes parentheses in the required palindromic order (Braitenberg and Pulvermüller, 1992; Pulvermüller, 1993).

References

Abeles M (1982) Local cortical circuits. Springer, Berlin Heidelberg New York
Abeles M (1991) Corticonics: Neural circuits of the cerebral cortex. Cambridge University Press, Cambridge
Abeles M, Prut Y (1996) Spatio-temporal firing patterns in the frontal cortex of behaving monkeys. J Physiol 90:249–250
Abercrombie M (1946) Estimation of nuclear population from microtome sections. Anat Rec 94:239–247
Aertsen AMHJ, Diesmann M, Gewaltig MO (1996) Propagation of synchronous spiking activity in feedforward neural networks. J Physiol 90:243–247
Aertsen AMHJ, Gerstein GL (1985) Evaluation of neuronal connectivity: sensitivity of cross-correlation. Brain Res 340:341–354
Amir Y, Harel M, Malach R (1993) Cortical hierarchy reflected in the organization of intrinsic connections in macaque monkey visual cortex. J Comp Neurol 334:19–46
Amit DJ (1989) Modelling brain function. The world of attractor neural networks. Cambridge University Press, Cambridge
Amit DJ, Brunel N (1997) Model of global spontaneous activity and local structured activity during delay periods in the cerebral cortex. Cerebr Cortex 7:237–252
Anokhin AP, Birbaumer N, Lutzenberger W, Nikolaev A, Vogel F (1996) Age increases brain complexity. Electroencephalography a Clinical Neurophysiology 99:63–68
Apfelbach R (1986) Imprinting on prey odours in ferrets (*mustela putorius f. furo l.*) and its neural correlates. Behav Processes 12:363–381
Apfelbach R, Weiler E (1985) Olfactory deprivation enhances normal spine loss in the olfactory bulb of developing ferrets. Neurosci Lett 62:169–173
Bär TH (1977) Wirkung chronischer Hypoxie auf die postnatale Synaptogenese im Occipitalcortex der Ratte. Verh Anat Ges 71:915–924
Bailey CH, Kandel ER (1993) Structural changes accompanying memory storage. Ann Rev Physiol 55:397–426
Bailey P, von Bonin G (1951) The isocortex of man. Univ of Illinois Press, Indiana.
Baillarger JGF (1840) Recherches sur la structure de la couche corticale des circonvolutions du cerveau. Mém. Acad. roy. d. méd. 8:149–183
Beaulieu C, Colonnier M (1985) A laminar analysis of the number of round-asymmetrical and flat-symmetrical synapses on spines, dendritic trunks, and cell bodies in area 17 of the cat. J Comp Neurol 231:180–189
Bienenstock E (1995) A model of neocortex. Network-computation in neural systems 6:179–224
Bienenstock E (1996) Composition. In: Aertsen A, Braitenberg B (eds) Brain theory. Biological basis and computational principles. Elsevier, Amsterdam, pp 269–300
Blackstad TW (1965) Mapping of experimental axon degeneration by electron microscopy of Golgi preparations. Z Zellforsch 67:819–834
Blackstad TW (1981) Tract tracing by electron microscopy of Golgi preparations. In: Heimer L, Robards MJ (eds) Neuroanatomical tract-tracing methods. Plenum, New York, pp 407–440
Blasdel GG, Salama G (1986) Voltage-sensitive dyes reveal a modular organization in monkey striate cortex. Nature 321:579–585

Blinkov SM, Glezer II (1968) Das Zentralnervensystem in Zahlen und Tabellen. VEB Gustav Fischer, Jena
Bliss TVP, Gardner-Medwin AR (1973) Long-lasting potentiation of synaptic transmission in the dentate area of the unanaesthetized rabbit following stimulation of the perforant path. J Physiol Lond 232:357–374
Bliss TVP, Lømo T (1973) Long-lasting potentiation of synaptic transmission in the dentate area of the anaesthetized rabbit following stimulation of the perforant path. J Physiol Lond 232:331–356
Bloom FE, Aghajanian GK (1966) Cytochemistry of synapses: selective staining for electron microscopy. Science 154:1575–1577
Bloom FE, Aghajanian GK (1968) Fine structural and cytochemical analysis of the staining of synaptic junctions with phosphotungstic acid. J Ultrastruct Res 22:361–375
Blue ME, Parnavelas JG (1983) The formation and maturation of synapses in the visual cortex of the rat. II. Quantitative analysis. J Neurocyt 12:697–712
Bok ST (1959) Histonomy of the cerebral cortex. Elsevier, Amsterdam
Bonhoeffer T, Staiger V, Aertsen AMHJ (1989) Synaptic plasticity in rat hippocampal slice cultures:local "Hebbian" conjunction of pre- and postsynaptic stimulation leads to distributed synaptic enhancement. Proc Natl Acad Sci USA 86:8113–8117
Braak H, Braak E (1976) The pyramidal cells of Betz within the cingulate cortex and precentral gigantopyramidal field in the human brain. Cell Tiss Res 172:103–119
Braak H, Braak E (1986) Ratio of pyramidal cells versus non-pyramidal cells in the human frontal isocortex and changes in ratio with ageing and Alzheimer's disease. In: Swaab DF, Fliers E, Mirmiran M, Van Gool WA, Van Haaren F (eds) Progress in brain research. Elsevier, Amsterdam, pp 185–212
Braendgaard H, Evans SM, Howard CV, Gundersen HJG (1990) The total number of neurons in the human neocortex unbiasedly estimated using optical disectors. J Microscopy 157:285–304
Braitenberg V (1962) A note on myeloarchitectonics. J. Comp. Neur. 118:141–156
Braitenberg V (1974a) Thoughts on the cerebral cortex. J Theor Biol 46:421–447
Braitenberg V (1974b) On the representation of objects and their relations in the brain. In: Conrad M, Güttinger W, Dal Cin M (eds) Lecture notes in biomathematics 4, Physics and mathematics of the nervous system. Springer, Berlin Heidelberg New York, pp 290–298
Braitenberg V (1977) On the texture of brains. Springer, Berlin Heidelberg New York
Braitenberg V (1978a) Cell assemblies in the cerebral cortex. In: Heim R, Palm G (eds) Lecture notes in biomathematics (21). Theoretical approaches to complex systems. Springer, Berlin Heidelberg New York, pp 171–188
Braitenberg V (1978b) Cortical architectonics: general and areal. In: Brazier MAB, Petsche H (eds) Architectonics of the cerebral cortex. Raven, New York, pp 443–465
Braitenberg V (1980) Alcune considerazioni sui meccanismi cerebrali del linguaggio. In: Braga G, Braitenberg V, Cipolli C, Coseriu E, Crespi – Reghizzi S, Mehler J, Titone R (eds), Franco Angeli Editore, Milano, pp 96–108
Braitenberg V (1981) Anatomical basis for divergence, convergence and integration in the cerebral cortex. Adv Physiol Sci 16:411–419
Braitenberg V (1983) Explanation of orientation columns in terms of a homogeneous network of neurons in the visual cortex. Soc Neurosci Abstr 9:474
Braitenberg V (1984) Vehicles:Experiments in synthetic psychology. MIT Press, Cambridge Mass.
Braitenberg V (1985) Charting the visual cortex. In: Peters A, Jones EG (eds) Cerebral cortex, Vol. 3, Plenum, New York, pp 379–414
Braitenberg V (1986) Two views of the cerebral cortex. In: Palm G, Aertsen A (eds) Brain theory. Springer, Berlin Heidelberg New York, pp 81–96

Braitenberg V (1992) How ideas survive evidence to the contrary: a comment on data display and modelling. In: Aertsen A, Braitenberg V (eds) Information processing in the cortex. Springer, Berlin Heidelberg New York, pp 447–450

Braitenberg V (1996) Il gusto della lingua:Meccanismi cerebrali e strutture grammaticali, Alpha & Beta, Merano

Braitenberg V, Braitenberg C (1979) Geometry of orientation columns in the visual cortex. Biol Cybern 33:179–186

Braitenberg V, Lauria F (1960) Toward a mathematical description of the grey substance of nervous systems. Nuovo Cimento 18:1135–1151

Braitenberg V, Pulvermüller F (1992) Entwurf einer neurologischen Theorie der Sprache. Naturwiss 79:103–117

Braitenberg V, Schüz A (1983) Some anatomical comments on the hippocampus. In: Seifert W (ed) Neurobiology of the Hippocampus. Academic Press, London, pp 21–37

Braitenberg V, Schüz A (1991) Anatomy of the cortex. Statistics and geometry. Springer, Berlin Heidelberg New York

Braitenberg V, Schüz A (1992) Basic features of cortical connectivity and some consideration on language. In: Wind J et al (eds):Language origin. A multidisciplinary approach, pp 89–102

Brodmann K (1909) Vergleichende Lokalisationslehre der Großhirnrinde. Barth, Leipzig

Buell SJ, Coleman PD (1981) Quantitative evidence for selective dendritic growth in normal human aging but not in senile dementia. Brain Res 214:23–41

Byrne JH (1985) Neural and molecular mechanisms underlying information storage in aplysia:implications for learning and memory. TINS November 478–482

Cajal S Ramón y (1911) Histologie du systme nerveux de l'homme et des vertébrés, translated by L Azoulay. Consejo superior de investigaciones cientificas. Instituto Ramon y Cajal, Madrid, edn 1972

Calverley RKS, Jones DG (1990) Contributions of dendritic spines and perforated synapses to synaptic plasticity. Brain Res Rev 15:215–249

Casagrande VA, Kaas JH (1994) The afferent, intrinsic, and efferent connections of primary visual cortex in primates. In: Peters A, Rockland KS (eds) Cerebral cortex, Vol. 10, Plenum, New York, pp 201–259

Caviness VS (1975) Architectonic map of neocortex of the normal mouse. J Comp Neurol 164:247–264

Chapin JK, Lin CS (1990) The somatic sensory cortex of the rat. In: Kolb B, Tees RC (eds) The cerebral cortex of the rat. MIT Press, Cambridge Mass, pp 341–380

Clarke PGH (1993) An unbiased correction factor for cell counts in histological sections. J Neurosc. Meth. 49:133–140

Collingridge GL, Bliss TVP (1987) NMDA receptors – their role in long-term potentiation. TINS 10:288–293

Colonnier M (1964) The tangential organization of the visual cortex. J Anat Lond 98:327–344

Colonnier M (1968) Synaptic patterns on different cell types in the different laminae of the cat visual cortex. An electron microscope study. Brain Res 9:268–287

Colonnier M (1981) The electron-microscopic analysis of the neuronal organization of the cerebral cortex. In: Schmitt FO, Worden FC, Adelman G, Dennis SG (eds) The organization of the cerebral cortex. MIT Press, Cambridge, Mass, pp 125–152

Cragg BG (1967) The density of synapses and neurones in the motor and visual areas of the cerebral cortex. J Anat 101:639–654

Cragg BG (1975) The development of synapses in the visual system of the cat. J Comp Neurol 160:147–166

Das A, Gilbert CD (1997) Distortions of visuotopic map match orientation singularities in primary visual cortex. Nature 387:594–598

DeFelipe J, Conley M, Jones EG (1986) Long-range focal collateralization of axons arising from corticocortical cells in monkey sensory-motor cortex. J Neurosci 6 (12):3749–3766
DeFelipe J, Jones EG (1988) Cajal on the cerebral cortex. An annotated translation of the complete writings. Oxford University Press, New York
Deuchars J, Thomson AM (1995) Innervation of burst firing spiny interneurons by pyramidal cells in deep layers of rat somatomotor cortex: paired intracellular recordings with biocytin filling. Neuroscience 69:739–755
De Vries J (1912) Über die Zytoarchitektonik der Grosshirnrinde der Maus und über die Beziehungen der einzelnen Zellschichten zum Corpus callosum aufgrund von experimentellen Läsionen. Folia Neurobiol 6:288–322
Drenhaus U, Schingnitz G, Dorka M (1986) A method for a quantitative determination of changes in tissue volume as a result of perfusion fixation. Anat Anz Jena 161:327–332
Drooglever Fortuyn AB (1914) On the cell-lamination of the cerebral cortex in some rodents. Arch Neurol 6:221–354
Economo von C, Koskinas GN (1925) Die Cytoarchitectonic der Hirnrinde des erwachsenen Menschen. Springer, Wien Berlin
Erwin E, Obermayer K, Schulten K (1995) Models of orientation and ocular dominance columns in the visual cortex:a critical comparison. Neur Comput 7:425–468
Fairén A, Valverde F (1980) A specialized type of neuron in the visual cortex of the cat: a Golgi and electron microscope study of chandelier cells. J Comp Neurol 194:761–779
Fairén A, DeFelipe J, Martínez-Ruiz R (1981) The Golgi-EM procedure: a tool to study neocortical interneurons. In: Eleventh International Congress of Anatomy: glial and neuronal cell biology. Liss, New York, pp 291–301
Fairén A, DeFelipe J, Regidor J (1984) Nonpyramidal neurons: general account. In: Peters A, Jones EG (eds) cerebral cortex, vol 1. Cellular components of the cerebral cortex. Plenum, New York, pp 201–253
Fairén A, Peters A, Saldanha J (1977) A new procedure for examining Golgi-impregnated neurons by light and electron microscopy. J Neurocytol 6:311–337
Feldman ML (1984) Morphology of the neocortical pyramidal neuron. In: Peters A, Jones EG (eds) Cerebral cortex, vol. 1. Cellular components of the cerebral cortex. Plenum, New York, pp 123–200
Feldman ML, Peters A (1979) A technique for estimating total spine numbers on Golgi-impregnated dendrites. J Comp Neurol 188:527–542
Felleman DJ, Van Essen D (1991) Distributed hierarchical processing in the primate cerebral cortex. Cerebral Cortex 1:1–47
Fifková E, Anderson CL (1981) Stimulation-induced changes in dimensions of stalks of dendritic spines in the dentate molecular layer. Exp Neurol 74:621–627
Fifková E, Delay RJ (1982) Cytoplasmic actin in neuronal processes as a possible mediator of synaptic plasticity. J Cell Biol 95:345–350
Fitzpatrick D, Lund JS, Schmechel DE, Towles AC (1987) Distribution of GABAergic neurons and axon terminals in the macaque striate cortex. J Comp Neur 264:73–91
Foh E, Haug H, König M, Rast A (1973) Quantitative Bestimmung zum feineren Aufbau der Sehrinde der Katze, zugleich ein methodischer Beitrag zur Messung des Neuropils. Microsc Acta 75:148–168
Garey LJ, Pettigrew JD (1974) Ultrastructural changes in kitten visual cortex after environmental modification. Brain Res 66:165–172
Gennari F (1782) De peculiari structura cerebri nonnullisque eius morbus. Parma
Gerstein GL (1962) Mathematical models for the all-or-none activity of some neurons. IRE Transactions on Information Theory IT-8:137–143
Gerstein GL, Aertsen AMHJ (1985) Representation of cooperative firing among simultaneously recorded neurons. J Neurophysiol 54:1513–1528

Gilbert CD, Wiesel TN (1979) Morphology and intracortical projections of functionally characterized neurons in the cat visual cortex. Nature 280:120–125
Gilbert CD, Wiesel TN (1989) Columnar specificity of intrinsic horizontal and corticocortical connections in cat visual cortex. J Neurosc 9:2432–2442
Globus A (1975) Brain morphology as a function of presynaptic morphology and activity. In: Riesen AH (ed) The developmental neuropsychology of sensory deprivation. Academic Press, New York, pp 9–91
Globus A, Scheibel AB (1967a) Pattern and field in cortical structure:the rabbit. J Comp Neurol 131:155–172
Globus A, Scheibel AB (1967b) The effect of visual deprivation on cortical neurons: a Golgi study. Exp Neurol 19:331–345
Götz KG (1987) Do "d-blob" and "l-blob" hypercolumns tessellate the monkey visual cortex? Biol Cybern 56:107–109
Golomb D, Rubin N, Sompolinsky H (1990) Willshaw model: associative memory with sparse coding and low firing rates. Phys Rev A 41:1843–1854
Gray EG (1959) Electron microscopy of synaptic contacts on dendrite spines of the cerebral cortex. Nature 183:1592–1593
Greenough WT, Chang FLF (1988) Plasticity of synapse structure and pattern in the cerebral cortex. In: Peters A, Jones EG (eds) Cerebral cortex, Vol. 7, Plenum, New York, pp 391–440
Greilich H (1984) Quantitative Analyse der cortico-corticalen Fernverbindungen bei der Maus. Thesis, University of Tübingen, FRG
Griesinger CB (1997) Molekulare Mechanismen struktureller Entwicklungsplastizität. Eine kombiniert elektrophysiologisch/anatomische Untersuchung. Dissertation an der Fakultät für Biologie der Universität Tübingen
Grinvald A, Lieke E, Frostig RD, Gilbert CD, Wiesel TN (1986) Functional architecture of cortex revealed by optical imaging of intrinsic signals. Nature 324:361–364
Grinvald A, Malach R (1994) Functional architecture and connection rules in primary visual cortex of macaque monkey. In: Albowitz B, Albus K, Kuhnt U, Nothdurft HChr, Wahle E (eds) Structural and functional organization of the neocortex: Proceedings of a symposium in the memory of Otto D Creutzfeld, May 1993. Springer, Berlin Heidelberg New York, pp 291–304
Groen van T, Vogt BA, Wyss JM (1993) Interconnections between the thalamus and the retrosplenial cortex in the rodent brain. In: Vogt BA, Gabriel M (eds) Neurobiology of the cingulate cortex and limbic thalamus. Birkhäuser, Boston, pp 123–150
Hartenstein V, Innocenti GM (1981) The arborization of single callosal axons in the mouse cerebral cortex. Neurosci Lett 23:19–24
Haug H (1979 a) Nervous tissue. In: Weibel ER (ed) Stereological methods, vol. 1. Practical methods for biological morphometry. Academic Press, London, pp 311–322
Haug H (1979 b) The evaluation of cell-densities and of nerve-cell-size distribution by stereological procedures in a layered tissue (cortex cerebri). Microsc Acta 82:147–161
Haug H (1986) History of neuromorphometry. In: Uylings HBM, Verwer RWH, Van Pelt J (eds) Morphometry and stereology in neurosciences. Special Issue J of Neurosc Meth 18,1–2:1–17
Hebb DO (1949) Organization of behavior. A neuropsychological theory, 2nd edn (1961). Wiley & Sons, New York
Hedlich A, Lüth HJ, Werner L, Bär B, Hanisch U, Winkelmann E (1990) GABAerge NADPH-diaphorase-positive Martinottizellen im visuellen Cortex der Ratte. J Hirnforsch 31:681–687
Hedreen JC (1981) High proportion of callosal-axon neurons in four areas of rat cerebral cortex. Anat Rec 199:108–109
Hellwig B (1992) Dichte und Verteilung präsynaptischer Boutons. Ein Beitrag zur Synaptologie der Grosshirnrinde. Thesis at the Faculty of Medicine, University of Tübingen, FRG

Hellwig B (1993) How the myelin picture of the human cerebral cortex can be computed from cytoarchitectural data. A bridge between von Economo and Vogt. J Hirnforsch 34, 3:387–402
Hellwig B (1994) In search of the structure of cell assemblies: how the connectivity between layer 2/3 pyramidal neurons in the rat visual cortex depends on the distance between their cell bodies. In: Elsner N, Breer H (eds) Göttingen neurobiology report 1994:proceedings of the 22nd Göttingen neurobiology conference 1994, vol. 2, p 564
Hellwig B (1995) A detailed analysis of the network of pyramidal neurons in layers 2/3 of the rat visual cortex. In: Elsner N, Menzel R (eds) Göttingen neurobiology report 1995: proceedings of the 23rd Göttingen neurobiology conference 1995, vol. 2, p 540
Hellwig B (in preparation) A quantitative analysis of the local connectivity between pyramidal neurons in layers 2/3 of the rat visual cortex.
Hellwig B, Schüz A, Aertsen A (1994) Synapses on axon collaterals of pyramidal cells are spaced at random intervals: a Golgi study in the mouse cerebral cortex. Biol Cybern 71:1–12
Hendry SHC, Hockfield S, Jones EG, McKay R (1984) Monoclonal antibody that identifies subsets of neurons in the central visual system of monkey and cat. Nature 307:267–269
Hendry SHC, Jones EG, Schwark HD, Yan J (1987) Numbers and proportions of GABA-immunoreactive neurons in different areas of monkey cerebral cortex. J Neurosci 7:1503–1519
Herrmann C, Schulz E (1978) Quantitative Untersuchungen an Sternzellen im Bereich der cingulären Rinde der Ratte. J Hirnforsch 19:519–531
Heumann D, Leuba G, Rabinowicz T (1977) Postnatal development of the mouse cerebral neocortex. II Quantitative cytoarchitectonics of visual and auditory areas. J Hirnforsch 18:483–500
Hopf A (1954) Die Myeloarchitektonik des Isocortex temporalis beim Menschen. J hirnforsch 1:208–279
Hopf A (1955) Über die Verteilung myeloarchitectonischer Merkmale in der isokorticalen Schläfenlappenrinde beim Menschen. J Hirnforsch 2:36–54
Hopfield JJ (1982) Neural networks and physical systems with emergent collective computational abilities. Proc Natl Acad Sci USA 79:2554–2558
Horner CH (1993) Plasticity of the dendritic spine. Progr Neurobiol 41:281–321
Horton JC, Hubel DH (1981) Regular patchy distribution of cytochrome oxidase staining in primary visual cortex of macaque monkey. Nature 292:762–764
Hubel DH, Wiesel TN (1965) Binocular interaction in striate cortex of kittens reared with artificial squint. J Neurophysiol 28:1041–1059
Hubel DH, Wiesel TN (1977) Functional architecture of macaque monkey visual cortex. Proc R Soc Lond Ser B 198:1–59
Humphrey NK, Hendrickson AE (1980) Radial zones of high metabolic activity in squirrel monkey striate cortex. Soc Neurosci Abstr 6:315
Huttenlocher PR, de Courten C, Garey LJ, van der Loos H (1982) Synaptogenesis in human visual cortex - evidence for synapse elimination during normal development. Neurosci Lett 33:247–252
Isenschmid R (1911) Zur Kenntnis der Grosshirnrinde der Maus. Abhandl K Preuss Akad
Isseroff A, Schwartz ML, Dekker JJ, Goldman-Rakic PS (1984) Columnas Organization of callosal and associational projections from rat frontal cortex. Brain Res 293:213–223
Jacobson S (1967) Dimensions of the dendritic spine in the sensorimotor cortex of rat, cat, squirrel monkey, and man. J Comp Neurol 129:49–58
Jacobson S, Trojanowski JQ (1974) The cells of origin of the corpus callosum in rat, cat and rhesus monkey. Brain Res 74:149–155

Jakobson, R (1941) Kindersprache, Aphasie und allgemeine Lautgesetze. Almgvist and Wiksell, Uppsala
Jerison HJ (1973) Evolution of the brain and intelligence. Academic Press, New York, San Francisco
Jerison HJ (1987) Brain size. In: Adelmann G (ed) Encyclopedia of neuroscience, Vol. 1. Birkhäuser, Boston, pp. 168–170
Jones EG (1984 a) Laminar distribution of cortical efferent cells. In: Peters A, Jones EG (eds) Cerebral cortex, vol. 1. Cellular components of the cerebral cortex. Plenum, New York, pp 521–553
Jones EG (1984 b) Neurogliaform or spiderweb cells. In: Peters A, Jones EG (eds) Cerebral cortex, vol 1. Cellular components of the cerebral cortex. Plenum, New York, pp 409–418
Jones EG, Hendry SHC (1984) Basket cells. In: Peters A, Jones EG (eds) Cerebral cortex, vol. 1. Cellular components of the cerebral cortex. Plenum, New York, pp 309–334
Jones EG, Hendry SHC, DeFelipe J, Benson DL (1994) GABA neurons and their role in activity-dependent plasticity of adult primate visual cortex. In: Peters A, Rockland KS (eds) Cerebral cortex, Vol. 10, Plenum, New York, pp 61–140
Jonson KM, Lyle JG, Edwards MJ, Penny RHC (1975) Problems in behavioural research with the guinea pig: a selective review. Anim Behav 23:632–639
Kawaguchi Y, Kubota Y (1996) Physiological and morphological identification of somatostatin- or vasoactive intestinal polypeptide-containing cells among GABAergic cell subtypes in rat frontal cortex. J Neurosci 16(8):2701–2715
Kennedy MB (1989) Regulation of neuronal function by calcium. TINS 12:417–420
Kirschfeld K, Feiler R, Wolf-Oberhollenzer F (1996) Cortical oscillations and the origin of express saccades. Proc R Soc Lond B 263:459–468
Kisvarday ZF, Martin KAC, Freund TF, Magloczky ZF, Whitteridge D, Somogyi P (1986) Synaptic targets of HRP-filled layer III pyramidal cells in the cat striate cortex. Exp Brain Res 64:541–552
Klimesch W (1996) Memory processes, brain oscillations and EEG synchronization. Internat J Psychophysiol 24:61–100
Kohonen T (1977) Associative memory. Springer, Berlin Heidelberg New York
König P, Engel AK, Singer W (1995) Relation between oscillatory activity and long-range synchronization in cat visual-cortex. Proc Natl Acad Sci USA 92:290–294
König P, Janosch B, SchillenTB (1992) Stimulus-dependent assembly formation of oscillatory responses. *Neural computation 4:666–681
Konigsmark BW (1970) Methods for the counting of neurons. In: Nauta WJH, Ebbesson SOE (eds) Contemporary research methods in neuroanatomy. Springer, Berlin Heidelberg New York, pp 315–340
Krieg WJS (1946) Connections of the cerebral cortex. I. The albino rat. A topography of the cortical areas. J Comp Neurol 84:277–324
Krone G, Mallot H, Palm G, Schüz A (1986) Spatiotemporal receptive fields:a dynamical model derived from cortical architectonics. Proc R Soc Lond B:421–444
Lashley KS (1951) The problem of serial order in behaviour. In: Jeffress LA (ed) Cerebral mechanisms in behaviour. Hafner, New York, pp 112–136
Lashley KS, Clark G (1946) The cytoarchitecture of the cerebral cortex of Ateles: a critical examination of architectonic studies. J Comp Neurol 85:223–306
LeVay S (1973) Synaptic patterns in the visual cortex of the cat and monkey. Electron microscopy of Golgi preparations. J Comp Neurol 150:53–86
Liley DTJ, Wright JJ (1994) Intracortical connectivity of pyramidal and stellate cells: estimates of synaptic densities and coupling symmetry. Network 5:175–189
Löwel S (1992) Blobs or slabs – is that the question? In: Aertsen A, Braitenberg V (eds) Information processing in the cortex. Springer, Berlin Heidelberg New York, pp 441–445

Löwel S, Freeman B, Singer W (1987) Topographic organization of the orientation column system in large flat-mounts of the cat visual cortex: a 2-deoxyglucose study. J Comp Neurol 255:401–415

Lorente de Nó R (1922) La corteza cerebral del ratón. Trab Lab Invest Biol Madrid 20:41–78

Lorente de Nó R (1938) Cerebral cortex: architecture, intracortical connections, motor projections. In: Fulton JF (ed) Physiology of the nervous system. 2nd edn, 1943. Oxford University Press, London, pp 274–313

Lübke J, Markram H, Frotscher M, Sakmann B (1996) Frequency and dendritic distribution of autapses established by layer 5 pyramidal neurons in the developing rat neocortex: comparisons with synaptic innervation of adjacent neurons of the same class. J Neurosci 16(10):3209–3218

Lund JS (1973) Organization of neurons in the visual cortex, area 17, of the monkey (*Macaca mulatta*). J Comp Neurol 147:455–496

Lund JS (1984) Spiny stellate neurons. In: Peters A, Jones EG (eds) Cerebral cortex, vol. 1. Cellular components of the cerebral cortex. Plenum, New York, pp 255–308

Lund, JS, Hendrickson, AE, Ogren, MP, Tobin, EA (1981) Anatomical organization of primate visual cortex area V II. J Comp Neurol 202:19–45

Lund JS, Wu Q, Hadingham PT, Levitt JB (1995) Cells and circuits contributing to functional properties in area VI of macaque monkey cerebral cortex:bases for neuroanatomically realistic models. J Anat 187:563–581

Malach R, Amir Y, Harel M, Grinvald A (1993) Relationship between intrinsic connections and functional architecture revealed by optical imaging and *in vivo* targeted biocytin injections in primate striate cortex. Proc Natl Acad Sci USA 90:10469–10473

Malenka RC, Kauer JA, Perkel DJ, Nicoll RA (1989) The impact of postsynaptic calcium on synaptic transmission – its role in long-term potentiation. TINS 12:444–450

Marin-Padilla M (1967) Number and distribution of the layer V pyramidal cells in man. J Comp Neurol 131:475–490

Marin-Padilla M (1969) Origin of the pericellular baskets of the pyramidal cells of the human motor cortex: A Golgi study. Brain Res 14:633–646

Marin-Padilla M (1972) Structural organization of the cerebral cortex (motor area) in human chromosomal aberrations. A Golgi study. Brain Res 44:625–629

Marin-Padilla M (1984) Neurons of layer I:a developmental analysis. In: Peters A, Jones EG (eds) Cerebral cortex, vol. 1. Cellular components of the cerebral cortex. Plenum, New York, pp 447–478

Markram H, Lübke, J, Frotscher M, Sakmann B (1997) Regulation of synaptic efficacy by coincidence of postsynaptic APs and EPSPs. Science 275:213–215

Martinotti C (1889) Contributo allo studio della corteccia cerebrale, ed all'origine centrale dei nervi. Ann Freniatr Sci Affini 1:14–381

Mason A, Nicoll A, Stratford K (1991) Synaptic transmission between individual pyramidal neurons of the rat visual cortex *in vitro*. J Neurosci 11 (1):72–84

Mayhew TM (1979) Stereological approach to the study of synapse morphometry with particular regard to estimating number in volume and on a surface. J Neurocytol 8:121–138

McDonald AJ (1992) Cell types and intrinsic connections of the amygdala. In: Aggleton JP (ed), The Amygdala. Neurobiological aspects of emotion, memory, and mental dysfunction. Wiley-Liss, New York, Chichester, pp 67–96

McGuire BA, Hornung J-P, Gilbert CD, Wiesel TN (1984) Patterns of synaptic input to layer 4 of cat striate cortex. J Neurosci 4:3021–3033

Michalski A, Patzwaldt R, Schulz E (1979) Quantitative Analyse der Neuronenstruktur der Regio retrosplenialis granularis der Ratte. J Hirnforsch 20:181–190

Miller M (1981) Maturation of rat visual cortex. I. A quantitative study of Golgi-impregnated pyramidal neurons. J Neurocytol 10:859–878

Miller MW, Vogt BA (1984) Direct connections of rat visual cortex with sensory, motor, and association cortices. J Comp Neurol 226:184–202
Miller R (1981) Meaning and purpose in the intact brain. Clarendon Press, Oxford
Miller R (1991) Cortico-hippocampal interplay and the representation of objects in the brain. Springer, Berlin
Miller R (1996) Neural assemblies and laminar interactions in the cerebral cortex. Biol Cybernetics 75:253–261
Miller R (1996) Cortico-thalamic interplay and the security of operation of neural assemblies and temporal chains in the cerebral cortex. Biol Cybern 75:263–275
Miller R, Wickens JR (1991) Corticostriatal cell assemblies in selective attention and in representation of predictable and controllable events. CINS 2,1:65–95
Mitra NL (1955) Quantitative analysis of cell types in mammalian neo-cortex. J Anat 89:467–483
Montero VM, Rojas A, Torrealba F (1973) Retinotopic organization of striate and peristriate visual cortex in the albino rat. Brain Res 53:197–201
Mouritzen Dam A (1979) Shrinkage of the brain during histological procedures with fixation in formaldehyde solutions of different concentrations. J Hirnforsch 20:115–119
Müller LJ, Verwer RWH, Nunes Cardozo B, Vrensen G (1984) Synaptic characteristics of identified pyramidal and multipolar non-pyramidal neurons in the visual cortex of young and adult rabbits. A quantitative Golgi-electron microscope study. J Neurosci 12:1071–1087
Müller MM, Bosch J, Elbert T, Kreiter A, Sosa MV, Sosa PV, Rockstroh B (1996) Visually induced gamma-band responses in human electroencephalographic activity – a link to animal studies. Experimental Brain Research 112:96–102
O'Kusky J, Colonnier M (1982) A laminar analysis of the number of neurons, glia, and synapses in the visual cortex (area 17) of adult macaque monkeys. J Comp Neurol 210:278–290
Palay SL, Chan-Palay V (1974) Cerebellar cortex. Springer, Berlin Heidelberg New York
Palm G (1982) Neural assemblies. An alternative approach to artificial intelligence. Springer, Berlin Heidelberg New York
Palm G (1986) Associative networks and cell assemblies. In: Palm G, Aertsen A (eds) Brain Theory. Springer, Berlin Heidelberg New York, pp 211–228
Palm G (1993) On the internal structure of cell assemblies. In: Aertsen A (ed) Brain theory. Spatio-temporal aspects of brain function. Elsevier, Amsterdam, pp 261–270
Palm G, Braitenberg V (1979) Tentative contributions of neuroanatomy to nerve net theories. In: Trappl R, Klir GJ, Ricciardi L (eds) Progress in cybernetics and systems research. Wiley & Sons, New York, pp 369–374
Pandya DN, Yeterian EH (1985) Architecture and connections of cortical association areas. In: Peters A, Jones EG (eds) Cerebral cortex. Association and auditory cortices. Plenum, New York, pp 3–61
Pantev C (1995) Evoked and induced gamma-band activity of the human cortex. Brain Topography 7:321–330
Parnavelas JG (1984) Physiological properties of identified neurons. In: Jones EG, Peters A (eds) Cerebral cortex, vol. 2. Functional properties of cortical cells. Plenum, New York, pp 205–239
Parnavelas JG, Burne RA, Lin C-S (1983) Distribution and morphology of functionally identified neurons in the visual cortex of the rat. Brain Res 261:21–29
Peters A (1979) Thalamic input to the cerebral cortex. TINS July:183–185
Peters A (1984a) Bipolar cells. In: Peters A, Jones EG (eds) Cerebral cortex, vol. 1. Cellular components of the cerebral cortex. Plenum, New York, pp 381–407
Peters A (1984b) Chandelier Cells. In: Peters A, Jones EG (eds) Cerebral cortex, vol. 1. Cellular components of the cerebral cortex. Plenum, New York, pp 361–380

Peters A (1987) Number of neurons and synapses in primary visual cortex. In: Jones EG, Peters A (eds) Cerebral cortex, vol. 6. Further aspects of cortical function including hippocampus. Plenum, New York, pp 267–294

Peters A, Feldman ML (1976) The projection of the lateral geniculate nucleus to area 17 of the rat cerebral cortex. I. General description. J Neurocytol 5:63–84

Peters A, Feldman ML (1977) The projection of the lateral geniculate nucleus to area 17 of the rat cerebral cortex. IV. Terminations upon spiny dendrites. J Neurocytol 6:669–689

Peters A, Jones EG (1984a) Cerebral cortex, vol. 1. Cellular components of the cerebral cortex. Plenum, New York

Peters A, Jones EG (1984b) Classification of cortical neurons. In: Peters A, Jones EG (eds) Cerebral cortex, vol. 1. Cellular components of the cerebral cortex. Plenum, New York, pp 107–121

Peters A, Kaiserman-Abramof IR (1970) The small pyramidal neuron of the rat cerebral cortex. The perikaryon, dendrites and spines. Am J Anat 127:321–355

Peters A, Kara DA (1985) The neuronal composition of area 17 of rat visual cortex. I. The pyramidal cells. J Comp Neurol 234:218–241

Peters A, Kimerer LM (1981) Bipolar neurons in rat visual cortex. A combined Golgi-electron microscopic study. J Neurocytol 10:921–946

Peters A, Proskauer CC (1980) Synaptic relationships between a multipolar stellate cell and a pyramidal neuron in the rat visual cortex. A combined Golgi-electron microscope study. J Neurocyt 9:163–183

Peters A, Palay SL, Webster H de F (1970) The fine structure of the nervous system. The cells and their processes. Harper & Row, New York

Peters A, Proskauer CC, Ribak CE (1982) Chandelier cells in rat visual cortex. J Comp Neurol 206:397–416

Peters A, Saint Marie RL (1984) Smooth and sparsely spinous nonpyramidal cells forming local axonal plexuses. In: Peters A, Jones EG (eds) Cerebral cortex, vol. 1. Cellular components of the cerebral cortex. Plenum, New York, pp 419–445

Petersen MR, Prosen CA, Moody DB, Stebbins WC (1977) Operant conditioning in the guinea pig. J Exp Anal Behav 27:529–532

Poljakow GI (1973) Foundations of neuronal taxonomy in the human neocortex. Medizina, Moscow (in Russian)

Porter L, White EL (1986) Synaptic connections of callosal projection neurons in the vibrissal region of mouse primary motor cortex: an electron microscopic/horseradish peroxidase study. J Comp Neurol 253:303–314

Pulvermüller F (1993) On connecting syntax and the brain. In: Aertsen A (ed), Brain theory. Elsevier, Amsterdam, pp 131–145

Pulvermüller F (1995) Agrammatism: Behavioral description and neurobiological explanation. J Cognit Neurosci 7, 2:165–181

Pulvermüller F, Eulitz C, Pantev C, Mohr B, Feige B, Lutzenberger W, Elbert T, Birbaumer N (1996) High-frequency cortical responses reflect lexical processing. An MEG study. Electroencephalography and Clinical Neurophysiology 98:76–85

Reith A, Mayhew TM (1988) Stereology and morphometry in electron microscopy. Some problems and their solutions. Hemisphere, New York

Ringo JL (1991) Neuronal interconnections as a function of brain size. Brain Behav Evol 38:1–6

Rockel AJ, Hiorns RW, Powell TPS (1980) The basic uniformity in structure of the neocortex. Brain 103:221–244

Romeis B (1968) Mikroskopische Technik. Oldenbourg Verlag, München

Rose M (1919) Histologische Lokalisation der Grosshirnrinde bei kleinen Säugetieren (Rodentia, Insectivora, Chiroptera) J Psychol Neurol 19:389–479

Rose M (1929) Cytoarchitektonischer Atlas der Grosshirnrinde der Maus. J Psychol Neurol 40:1–32

Rosenquist AC (1985) Connections of visual cortical areas in the cat. In: Peters A, Jones EG (eds) Cerebral cortex, Vol. 3, Plenum, New York, pp 81–117
Rumelhart DE, McClelland JL and the PDP Research Group (1986) Parallel distributed processing. Explorations in the microstructure of cognition. Vol. 1: Foundations. MIT Press, Cambridge, Mass
Rusakov DA, Stewart MG (1995) Quantification of dendritic spine populations using image analysis and a tilting disector. J Neurosci Meth 60:11–21
Sanides F (1962) Die Architektonik des menschlichen Stirnhirns. In: Müller M, Spath H, Vogel P (eds), Monographien aus dem Gesamtgebiete der Neurologie und Psychiatrie, Vol. 98. Springer, Berlin, Göttingen
Scannell JW, Blakemore C, Young MP (1995) Analysis of connectivity in the cat cerebral cortex. J Neurosc 15(2):1463–1483
Schapiro S, Vukovich K, Globus A (1973) Effects of neonatal thyroxine and hydrocortison administration on the development of dendritic spines in the visual cortex of rats. Exp Neurol 40:286–296
Schober W, Winkelmann E (1975) Der visuelle Kortex der Ratte. Cytoarchitektonik und stereotaktische Parameter. Z Mikrosk Anat Forsch Leipz 89:431–446
Schüz A (1976) Pyramidal cells with different densities of dendritic spines in the cortex of the mouse. Z Naturforsch 31c:319–323
Schüz A (1978) Some facts and hypotheses concerning dendritic spines and learning. In: Brazier MAB, Petsche H (eds) Architectonics of the cerebral cortex. Raven, New York, pp 129–135
Schüz A (1981a) Pränatale Reifung und postnatale Veränderungen im Cortex des Meerschweinchens: mikroskopische Auswertung eines natürlichen Deprivationsexperimentes. I. Pränatale Reifung. J Hirnforsch 22:93–111
Schüz A (1981b) Pränatale Reifung und postnatale Veränderungen im Cortex des Meerschweinchens: mikroskopische Auswertung eines natürlichen Deprivationsexperimentes. II. Postnatale Veränderungen. J Hirnforsch 22:113–127
Schüz A (1986) Comparison between the dimensions of dendritic spines in the cerebral cortex of newborn and adult guinea pigs. J Comp Neurol 244:277–285
Schüz A (1989) Untersuchungen zur Verknüpfungsstruktur der Grosshirnrinde. Quantitative Studien am Cortex der Maus. Thesis, University of Tübingen, FRG
Schüz A, Demianenko GP (1995) Constancy and variability in cortical structure:a study on synapses and dendritic spines in hedgehog and monkey. J Brain Res 36:113–122
Schüz A, Dortenmann M (1987) Synaptic density on non-spiny dendrites in the cerebral cortex of the house mouse. A phosphotungstic acid study. J Hirnforsch 28:633–639
Schüz A, Hein FM (1984) Comparison between the developmental calendars of the cerebral and cerebellar cortices in a precocial and an altricial rodent. In: Bloedel J, Dichgans J, Precht W (eds) Cerebellar Functions. Springer, Berlin Heidelberg New York, pp 318–321
Schüz A, Münster A (1985) Synaptic density on the axonal tree of a pyramidal cell in the cortex of the mouse. Neuroscience 15:33–39
Schüz A, Palm G (1989) Density of neurons and synapses in the cerebral cortex of the mouse. J Comp Neurol 286:442–455
Schwartzkroin PA, Wester K (1975) Long-lasting facilitation of a synaptic potential following tetanization in the in vitro hippocampal slice. Brain Res 89:107–119
Schweizer M (1990) Zur Entwicklung der dendritischen Dornen im Cortex der Ratte. Eine elektronenmikroskopische Untersuchung. Thesis, University of Tübingen, FRG
Sholl DA (1956) The organization of the cerebral cortex. Wiley & Sons, New York
Sholl DA (1959) A note on the neuronal packing density in the cerebral cortex. J Anat 93:434–435

Sidman RL, Angevine JB, Pierce ET (1971) Atlas of the mouse brain and spinal cord. Harvard University Press, Cambridge, Mass
Somogyi P (1977) A specific "axo-axonal" interneuron in the visual cortex of the rat. Brain Res 136:345–350
Somogyi P (1978) The study of Golgi-stained cells and of experimental degeneration under the electron microscope: a direct method for the identification in the visual cortex of three successive links in a neuron chain. Neuroscience 3:167–180
Somogyi P, Cowey A (1981) Combined Golgi and electron microscopic study on the synapses formed by double bouquet cells in the visual cortex of the cat and monkey. J Comp Neurol 195:547–566
Somogyi P, Nunzi MG, Gorio A, Smith AD (1983) A new type of specific interneuron in the monkey hippocampus forming synapses exclusively with the axon initial segments of pyramidal cells. Brain Res 259:137–142
Staiger V (1984) Internal report, unpublished
Stephan H (1960) Methodische Studien über den quantitativen Vergleich architektonischer Struktureinheiten des menschlichen Gehirns. Z wiss Zool 164:1–2
Sterio DC (1984) The unbiased estimation of number and sizes of arbitrary particles using the disector. J Microsc 134:127–136
Stevens CF (1989) How cortical interconnectedness varies with network size. Neural Computation 1:473–479
Swanson LW, Köhler C (1986) Anatomical evidence for direct projections from the entorhinal area to the entire cortical mantle in the rat. J Neurosci 6:3010–3023
Swindale NV (1981) Dendritic spines only connect. TINS September 1981:240–241
Szentágothai J (1969) Architecture of the cerebral cortex. In: Jasper HH, Ward AA, Pope A (eds) Basic mechanisms of the epilepsies. Little & Brown, Boston, pp 13–28
Szentágothai J (1975) The "module-concept" in cerebral cortex architecture. Brain Res 95:475–496
Thomson AM, Girdlestone D, West DC (1988) Voltage-dependent currents prolog single-axon postsynaptic potentials in layer III pyramidal neurons in rat neocortical slices. J Neurophysiol 60:1896–1906
Tower TB (1954) Structural and functional organization of mammalian cerebral cortex: The correlation of neruone density with brain size. J Comp Neurol 101:19–51
Tower DB, Elliott KAC (1952) Activity of acetylcholine system in cerebral cortex of various unanesthetized mammals. Am J Physiol 168:747–759
Tömböl T (1978) Comparative data on the Golgi architecture of interneurons of different cortical areas in cat and rabbit. In: Brazier MAB, Petsche HP (eds) Architectonics of the cerebral cortex. Int. Brain Res Organization Monograph Series, vol 3. Raven, New York, pp 59–76
Tömböl T (1984) Layer VI cells. In: Peters A, Jones EG (eds) Cerebral cortex, vol 1. Cellular components of the cerebral cortex. Plenum, New York, pp 479–519
Uchizono K (1965) Characteristics of excitatory and inhibitory synapses in the central nervous system of the cat. Nature 207:642–643
Underwood EE (1970) Quantitative stereology. Addison-Wesley, Readinger
Uylings HBM, Verwer RWH, Van Pelt J (1986) Morphometry and stereology in neurosciences. In: Uylings HBM, Verwer RWH, Van Pelt J (eds) J Neurosc Methods (Spec Iss). Elsevier, Amsterdam
Valverde F (1967) Apical dendritic spines of the visual cortex and light deprivation in the mouse. Exp Brain Res 3:337–352
Valverde F (1971) Rate and extent of recovery from dark rearing in the visual cortex of the mouse. Brain Res 33:1–11
Valverde F (1983) A comparative approach to neocortical organization based on the study of the brain of the hedgehog (*Erinaceus europaeus*) In: Grisolia S, Guerri C, Samson F, Norton S, Reinoso-Suárez F (eds) Ramón y Cajal's contribution to the neurosciences. Elsevier, Amsterdam, pp 149–170

Valverde F (1991) Aspects of phylogenetic variability of neocortical intrinsic organization. In: Finlay BL, Innocenti G, Scheich H (eds) The neocortex. Ontogeny and Phylogeny. Olenum, New York, pp 87–102
Van Essen DC (1997) A tension based theory of morphogenesis and compact wiring in the central-nervous-system. Nature 385:313–318
Varela FJ (1995) Resonant cell assemblies: A new approach to cognitive functions and neuronal synchrony. Biol Res 28:81–95
Vaughan DW, Peters A (1973) A three dimensional study of layer I of the rat parietal cortex. J Comp Neurol 149:355–370
Vogt BA (1985) Cingulate cortex. In: Peters A, Jones EG (eds) Cerebral cortex, vol. 4. Association and auditory cortices. Plenum, New York, pp 89–149
Vogt O (1910) Die meloarchitektonische Felderung des menschlichen Strinhirns. J Psychol Neurol 15:221–232
Vogt O (1911) Die Myeloarchitektonik des Isocortex parietalis. J Psychol Neurol 18:379–390
Vrensen G, de Groot D (1974) The effect of dark rearing and its recovery on synaptic terminals in the visual cortex of rabbits. A quantitative electron microscopic study. Brain Res 78:263–278
Wahle, P (1993) Differential regulation of substance P and somatostatin in Martinotti cells of the developing cat visual cortex. J Comp Neurol 329:519–538
Wallhäusser E, Scheich H (1987) Auditory imprinting leads to differential 2-deoxyglucose uptake and dendritic spine loss in the chick rostral forebrain. Dev Brain Res 31:29–44
Wang X, Merzenich MM, Sameshima K, Jenkins WM (1995) Modelling of hand representation in adult cortex determined by timing of tactile stimulation. Nature 387/6552:71–75
Waters RS, McCandlish CA, Li ChX (1995) Organization and development of the forepaw representation in forepaw barrel subfield in somatosensory cortex of rat. In: Jones EG, Diamond IT (eds) Cerebral cortex, Vol. 11, Plenum, New York, pp 77–122
Weibel ER (1969) Stereological principles for morphometry in electron microscopic cytology. Intern Rev Cytol 26:235–302
Welker C (1976) Receptive fields of barrels in the somatosensory neocortex of the rat. J Comp Neurol 166:173–190
Werner J (1986) Einbettungs- und Färbemethoden für das RWL-Medium. RWL Histotechnologie, Bruckmuehl
Werner L, Winkelmann E (1976) Untersuchungen zur Struktur der thalamo-kortikalen Projektionsneuronen und Interneuronen im Corpus geniculatum laterale pars dorsalis (Cgld) der Albinoratte nach unterschiedlicher histologischer Technik. Anat Anz Bd 139:142–157
Werner L, Hedlich A, Winkelmann E, Brauer K (1979) Versuch einer Identifizierung von Nervenzellen des visuellen Kortex der Ratte nach Nissl- und Golgi-Kopsch-Darstellung. J Hirnforsch 20:121–139
White EL (1978) Identified neurons in mouse SmI cortex which are postsynaptic to thalamocortical axon terminals: a combined Golgi-electron microscopic and degeneration study. J Comp Neurol 181:627–662
White EL (1979) Thalamocortical synaptic relations:a review with emphasis on the projections of specific thalamic nuclei to the primary sensory areas of the neocortex. Brain Res Rev 1:275–311
White EL (1981) Thalamocortical synaptic relations. In: Adelman G, Dennis SG, Schmitt FO, Worden FG (eds) The organization of the cerebral cortex. MIT Press, Cambridge, Mass, pp 153–161

White EL (1986) Terminations of thalamic afferents. In: Jones EG, Peters A (eds) Cerebral cortex, vol. 5. Sensory-motor areas and aspects of cortical connectivity. Plenum, New York, pp 271–289

White EL (1987) Thalamocortical interactions In: Adelman G (ed) Encyclopedia of neuroscience. Birkhäuser, Boston, pp 1202–1204

White EL (1989) Cortical circuits. Synaptic organization of the cerebral cortex. Structure, function, and theory. Birkhäuser, Boston

White EL, Hersch SM (1981) Thalamocortical synapses of pyramidal cells which project from Sml to Msl cortex in the mouse. J Comp Neurol 198:167–181

White EL, Hersch SM (1982) A quantitative study of thalamocortical and other synapses involving the apical dendrites of corticothalamic projection cells in mouse SmI cortex. J Neurocytol 11:137–157

White EL, Keller A (1987) Intrinsic circuitry involving the local axon collaterals of corticothalamic projection cells in mouse SmI cortex. J Comp Neurol 262:13–26

White EL, Rock MP (1979) Distribution of thalamic input to different dendrites of a spiny stellate cell. Neurosci Lett 15:115–119

White EL, Rock MP (1980) Three-dimensional aspects and synaptic relationships of a Golgi-impregnated spiny stellate cell reconstructed from serial thin sections. J Neurocytol 9:615–636

White EL, Rock MP (1981) A comparison of thalamocortical and other synaptic inputs to dendrites of two non-spiny neurons in a single barrel of mouse SmI cortex. J Comp Neurol 195:265–277

White EL, Benshalom G, Hersch SM (1984) Thalamocortical and other synapses of non-spiny multipolar cells in mouse SmI cortex. J Comp Neurol 229:311–320

Wickelgren WA (1992) Webs, cell assemblies, and chunking in neural nets. CINS 3, 1:1–53

Wickens JR (1988) Electrically coupled but chemically isolated synapses: dendritic spines and calcium in a rule for synaptic modification. Progr in Neurobiol 31:507–528

Wickens JR, Hyland B, Anson G (1994) Cortical assemblies: a possible mechanism for motor programs. J Motor Behav 26, 2:66–82

Wiesel TN, Hubel DH (1965) Comparison of the effects of unilateral and bilateral eye closure on cortical unit responses in kitten. J Neurophysiol 28:1029–1040

Wilson CJ, Groves PM, Kital ST, Linder JC (1983) Three-dimensional structure of dendritic spines in the rat neostriatum. J Neurosci 3:383–398

Winfield DA (1983) The postnatal development of synapses in the different laminae of the visual cortex in the normal kitten and in kittens with eyelid suture. Dev Brain Res 9:155–169

Winfield DA, Gatter KC, Powell TPS (1980) An electron microscopic study of the types and proportions of neurons in the cortex of the motor and visual areas of the cat and rat. Brain 103:245–258

Winfield DA, Brooke RNL, Sloper JJ, Powell TPS (1981) A combined Golgi-electron microscopic study of the synapses made by the proximal axon and recurrent collaterals of a pyramidal cell in the somatic sensory cortex of the monkey. Neuroscience 6:1217–1230

Winkelmann E, Brauer K, Werner L (1977) Untersuchungen zu Spineveränderungen der Lamina-V-Pyramidenzellen im visuellen Kortex junger und subadulter Laborratten nach Dunkelaufzucht und Zerstörung des corpus geniculatum laterale, pars dorsalis. J Hirnforsch 17:496–506

Wolff JR (1976) Quantitative analysis of topography and development of synapses in the visual cortex. Exp Brain Res Suppl 1:259–263

Wolff JR (1978) Ontogenetic aspects of cortical architecture:lamination. In: Brazier MAB, Petsche H (eds) Architectonics of the cerebral cortex. Raven, New York, pp 159–173

Wolff JR, Laskawi R, Spatz WB, Missler M (1995) Structural dynamics of synapses and synaptic components. Behav Brain Res 66:13–20
Woolsey TA (1967) Somatosensory, auditory and visual cortical areas of the mouse. John Hopkins Med J 121:91–112
Woolsey TA, van der Loos H (1970) The structural organization of layer IV in the somatosensory region (SI) of mouse cerebral cortex. Brain Res 17:205–242
Yorke CH, Caviness VS (1975) Interhemispheric neocortical connections of the corpus callosum in the normal mouse: a study based on anterograde and retrograde methods. J Comp Neurol 164:233–246
Young MP (1992) Objective analysis of the topological organization of the primate cortical visual system. Nature 358:152–154
Young MP, Scannell JW, Burns G (1995) The analysis of cortical connectivity. Springer, New York Berlin Heidelberg
Young MP, Scannell JW, Burns GAPC, Blakemore C (1994) Analysis of connectivity: neural systems in the cerebral cortex. Rev Neurosc 5:227–249

Subject Index

R

S

T

MIX
Papier aus verantwortungsvollen Quellen
Paper from responsible sources
FSC® C105338

If you have any concerns about our products,
you can contact us on
ProductSafety@springernature.com

In case Publisher is established outside the EU,
the EU authorized representative is:
Springer Nature Customer Service Center GmbH
Europaplatz 3, 69115 Heidelberg, Germany

Printed by Libri Plureos GmbH
in Hamburg, Germany